DATE DUE

The American Atom

The American Atom

A Documentary History of Nuclear Policies

from the Discovery of Fission to the Present

1939–1984

ROBERT C. WILLIAMS AND
PHILIP L. CANTELON,
Editors

upp

UNIVERSITY OF PENNSYLVANIA PRESS
Philadelphia

Copyright © 1984 by the University of Pennsylvania Press

Library of Congress Cataloging in Publication Data

Main entry under title:
The American atom.
   1. Atomic weapons—History—Sources.   2. Atomic energy—
History—Sources.   3. Atomic weapons and disarmament—
History—Sources.   4. Nuclear nonproliferation—History—
Sources.   5. United States—Military policy—Addresses,
essays, lectures.   6. Deterrence (Strategy)—History—20th
century—Sources.   I. Williams, Robert Chadwell, 1938–.
II. Cantelon, Philip L. (Philip Louis), 1940–.
U264.A44   1984   355.8′25119′0973   83-21829
ISBN 0-8122-7920-4
ISBN 0-8122-1169-3 (pbk.)

Printed in the United States of America

Fourth paperback printing, 1986

What concerns me is really not the technical problem. I am not sure the miserable thing will work, nor that it can be gotten to a target except by oxcart. It seems likely to me even further to worsen the unbalance of our war plans. What does worry me is that this thing appears to have caught the imagination, both of the Congressional and military people, as the answer to the problem posed by the Russians' advance. It would be folly to oppose the exploration of this weapon. We have always known it had to be done; and it does have to be done, though it appears to be singularly proof against any form of experimental approach. But that we become committed to it as the way to save the country and the peace appears to me full of dangers.

> J. Robert Oppenheimer to James Conant
> October 12, 1949

In 1950 our research group became part of a special institute. For the next eighteen years I found myself caught up in the rotation of a special world of military designers and inventors, special institutes, committees and learned councils, pilot plants and proving grounds. Every day I saw the huge material, intellectual, and nervous resources of thousands of people being poured into creating the means of total destruction, a force potentially capable of annihilating all human civilization.

> Andrei Sakharov
> 1974

Contents

VII   Deterrence   210

VIII   Arms Control   238

Introduction

Historical understanding is essential to an informed public that must live in a world of nuclear weapons and nuclear reactors because of decisions made years or even decades ago. The purpose of this volume is to tell the story of America's nuclear policy through its primary documents. Our hope is that a documentary collection will contribute historical evidence to a public concern with nuclear policy that is often more rhetorical than substantive.

Since 1979, when the SALT II treaty was withdrawn from Senate consideration and the Three Mile Island nuclear accident drew international attention, a public debate on nuclear policy without precedent since the mid-1950s has developed. A generation "born nuclear" and raised in a world of tens of thousands of nuclear warheads and hundreds of reactors is asking questions that had been the secret concern of only a few experts since 1945: should we proceed with the further development of nuclear weapons? Of nuclear reactors? Is the benefit worth the cost and the risk? Does deterrence work? Is arms control beneficial? Should America use nuclear weapons first? Are national technical means of verification reliable? Can you trust the Russians? Can a nuclear war be limited? Can it be won?

The early decisions to build nuclear and thermonuclear weapons were usually made in secret by nuclear and military specialists. Yet since 1979 decisions to install Cruise and Pershing II missiles in Europe, to build the MX missile, and to develop antimissile defense systems all have come under immediate concerned public scrutiny. Nuclear issues have assumed mythic proportions: nuclear war is the Apocalypse and Armageddon; nuclear power is a Faustian bargain; missiles are named after the gods Nike, Zeus, and Titan. What has often been missing is a knowledge of the facts.

After nearly fifty years the development of nuclear policy is certainly history, but its primary sources are scattered and often difficult to obtain, especially for the layman or student. All history involves selecting the significant documents in order to make sense of the past. Here we have focused on neither the technical calculations of nuclear war and its effects on familiar cities nor the endless proposals for a disarmed world. Rather, we have gathered together those documents that will best show how our present policies have evolved in a world that is neither engaged in a nuclear war nor disarmed, but living in the twilight zone of arms control and regulated nuclear power.

In selecting these historical documents we have applied three general tests: first, does the document help tell the story of the development of American nuclear policy in a nontechnical way; second, is the source primary rather than secondary, written by an actor in the drama rather than by a member of the audience; third, does the document provide coverage of the major chapters in the story?

In showing how the vision of a nuclear world preceded the reality by several decades, we eschewed the physics of Einstein and Bohr for the science fiction of H. G. Wells. We also included the correspondence of Leo Szilard, the dynamic Hungarian refugee who engineered the famous Einstein letter to Roosevelt and helped move America from neutrality to defense. Since the original contributions to the theory of the atom bomb were British, not American, we provided some of the early speculations of British nuclear physicists between 1940 and 1941. We also sought to show the reader how the nuclear age began in the minds of men living in a Europe at war.

The Manhattan Project was America's $2 billion secret project to build an atomic bomb. Many documents associated with the project have come to light only in recent years. In Section II we have used the letters of J. Robert Oppenheimer and the recently declassified minutes of policy committees to tell the story of how the bomb was designed and built and how the decision was made to drop the first uranium and plutonium devices on the Japanese cities of Hiroshima and Nagasaki in 1945.

How did a weapon of war become the key to a peacetime industry? In considering atomic energy after World War II, we focused in Section III on the legislative enabling acts that established the Atomic Energy Commission, the short-lived dream of international control of nuclear weapons under the Baruch Plan, and the "atoms for peace" program of President Dwight D. Eisenhower. By 1954 the highly classified work on nuclear weapons paralleled a new development of nuclear energy and power reactors. Knowledge was shared with both private industry and other countries. The fruits of this program are considered in the later section on nuclear power.

Until 1949 America was apparently secure in its monopoly of nuclear weapons. But the detonation of the first Soviet nuclear device, code named "Joe I," in August 1949 precipitated a frenzied race to construct the most awesome nuclear weapons of our time, based on the thermonuclear or fusion principle. In Section IV on the hydrogen bomb, we have tried to show how the Soviet threat impelled America toward a crash program to build the "super" weapon, opposed by J. Robert Oppenheimer and the Atomic Energy Commission's General Advisory Committee but approved by President Harry S. Truman in 1950. We have included the recently declassified minutes of the General Advisory Committee meeting of October 30, 1949, and Hans Bethe's account of the race for the H-bomb.

Why do scientists build nuclear weapons? In Section V on the Oppenheimer hearing of 1954, we examine a case study of science and conscience versus national security, a courtroom drama that drew the attention of both the public and concerned scientists, many of them Manhattan Project veterans. The withdrawal of Oppenheimer's security clearance after a decade of service to his country—based not on his supposed security risk but on his moral opposition to the H-bomb—seemed to many a dangerous insult to the entire scientific community. In selecting testimony and findings from "In the Matter of J. Robert Oppenheimer," we have tried to show how the Oppenheimer case brought to light the entire history of the Manhattan Project and dramatized the question of scientific conscience and loyalty to the state.

Nuclear testing was an essential part of the research and development of nuclear weapons, from the flats of Nevada to the sands of Enewetak, in the 1950s. Testing presented health hazards to a concerned public and raised fears of a nuclear war between the Soviet Union and the United States. For a time citizens sought security in a civil defense program of fallout shelters, but the dangers of worldwide nuclear testing continued. In 1963 a limited ban on atmospheric testing was effected by America and the Soviet Union, the first of a series of limited arms-control measures that were bilateral and outside the international framework of the United Nations. In Section VI we have selected government documents, statements reflecting public opinion, and treaties to indicate the level of concern over testing and its diminution after 1963.

What is deterrence? Does it work? How did a doctrine of "mutual assured destruction" come into being? In Section VII on the evolution of the central doctrine of the nuclear age—that the only rational use of nuclear weapons is to deter an aggressor by threatening retaliation if he strikes first—we have provided selections from RAND strategic thinkers Bernard Brodie and Herman Kahn along with statements from various administrations, from John Foster Dulles on "massive retaliation" in 1954 to President

Ronald Reagan's definition of deterrence in 1982. Deterrence is often more an accepted assumption than a stated doctrine, but these selections help give the reader an understanding of how deterrence has been articulated over three decades as the central rationale for American nuclear weapons policy.

Like deterrence, arms control is a policy developed over several decades and lies in a kind of no-man's-land between the paradise of disarmament and the hell of nuclear war. Both deterrence and arms control depend on the careful management of new and old weapons systems and careful communication with the Soviet Union. Can arms control work? In Section VIII we have tried to show how a series of limited and bilateral agreements between the United States and the Soviet Union following the Cuban missile crisis of 1962 attempted to manage communications and to control force levels of the two nuclear superpowers. Beginning with a listing of "broken arrows"—nuclear weapons accidents in the 1950s and 1960s—we show how the dream of international disarmament in the 1950s gave way to the more limited task of arms control through the hotline communications system between Moscow and Washington, the limiting of antiballistic missile systems in SALT I (1972), and the signing by President Jimmy Carter and Premier Leonid Brezhnev of the SALT II agreement in Vienna in 1979. Limited bilateral measures adopted in the common interest of the two superpowers continue to operate and form the basis of the ongoing Strategic Arms Reduction Talks (START) in Geneva in 1983. In our final selection William P. Clark of the National Security Council attempts to explain this historical process of arms control to the National Conference of Catholic Bishops.

Does nuclear power have a future? Experts and the public may disagree, but it clearly has a past. Beginning with the revision of the Atomic Energy Act in 1954, government and private industry became partners in a major effort to develop nuclear reactors as a source of electrical power in America. For a time in the 1960s the "great bandwagon market" that produced dozens of new light water reactors across the country convinced many that nuclear power could provide electricity that was "too cheap to meter." Gradually, however, the nuclear dream dissolved, in part because of increasing public concern with radiological dangers and increasing government regulation in the interest of public health and safety. The documents on nuclear power in Section IX show how the first major accident at the Fermi nuclear power plant in 1966 ("We almost lost Detroit") precipitated a series of new legal decisions and regulations that helped slow the growth of nuclear power and made it increasingly expensive. By the time of the Three Mile Island accident of 1979, orders for new nuclear plants had all but disappeared, and the public seemed more concerned with the risks than the benefits of nuclear power. Our final selection is not from the highly publicized Kemeny or Rogovin reports on the accident, but from a nuclear industry account written within a week of the accident. Like so many of the documents collected here, it provides a sense of immediate historical presence, rather than a qualified subsequent analysis.

The American atom is no longer a secret project but a historical legacy that we must try to understand and with which we must live. Perhaps no other issue of our times is as heavily weighted with history as is the nuclear issue. Without a clear understanding of the nature, purpose, and limits of arms control and regulated nuclear power we cannot understand the world of nuclear weapons and nuclear energy in which we live. We hope that this volume offers a starting point for understanding how we all came to live in the nuclear age, and for distinguishing between the human choices we have inherited and the ones we must still make.

We are greatly indebted to Washington University in St. Louis and to the Alfred P. Sloan Foundation for supporting this project, which is based on the experience of classroom teaching, as well as historical research. Many of these materials were developed in connection with a course on "Nuclear Energy and Contemporary History" offered in 1981–83 at Washington University through its Department of History, and under the program in Technology and the Liberal Arts. We are also grateful to WU Professors Arthur Wahl, Lindsay Helmholz, David Lipkin, and Sam Weissman for their candid recollections of Los Alamos days, and to Mr. William Cornelius of the Union Electric Company, Col. Robert Browning, and Professor George Hudson of Wittenberg University for sharing their nuclear expertise. Professor Martin Sherwin of Tufts University reviewed the entire manuscript and made a number of valuable suggestions. The result, of course, is our responsibility and not theirs, but we are grateful for their assistance.

We would also like to acknowledge the assistance of Antoinette Zemel, who cheerfully and efficiently typed the bulk of the manuscript, and to our editor, Malcolm Call. Venita Lake of Washington University has been a devoted and effective reader, editor, critic, and typist. We are also indebted to Ann Williams for her support and patience.

PHILIP L. CANTELON
ROBERT C. WILLIAMS
St. Louis, Missouri
Spring 1983

The Nuclear Age: Background and Visions

*T*he American atom had its origins in Europe at the turn of the century in the minds of physicists. Between 1900 and 1914 a veritable revolution occurred in our view of the material universe and its laws. The discovery of radioactivity in the late 1890s was an alchemist's dream; one chemical element could decay into another in a process of transmutation that gave off new sources of energy. A very small amount of matter, by Albert Einstein's 1905 calculations, could be transformed into a very large amount of energy proportionate to the square of the speed of light, a constant. X-rays saw through matter, which could be understood as waves rather than particles. The atom itself, that smallest and most discrete material object, appeared to be made up of still smaller particles, a nucleus surrounded by electrons.

The technological wonders of World War I—tanks, airplanes, flamethrowers, and submarines—demonstrated the power of science at war. But the discoveries of the 1920s and 1930s extended the revolution in physics still further. Small particles at high speeds could be located and measured only with a finite degree of uncertainty. Energy appeared to be released in discrete amounts as "light quanta," behaving sometimes as waves and at other times as particles. In 1932 experiments verified that the nucleus was composed of positively charged protons and uncharged neutrons, continuing the breakdown of the material world.

Even before World War I, when drinkers consumed radium cocktails that glowed in the dark of Paris and New York nightclubs, a few visionaries anticipated the coming of the nuclear age. In 1914 the British science-fiction writer and prophet of future technologies H. G. Wells wrote a book entitled The World Set Free, in which he envisaged by the 1950s a world of "atomic bombs" for war and nuclear reactors for peace.

In December 1938 two German physicists, Otto Hahn and Fritz Strassmann, working in a Berlin laboratory, discovered that the heavy element uranium upon bombardment by neutrons had split, or fissioned, into separate, lighter elements with a consequent production of both high energy and additional neutrons. It seemed possible that under proper control the additional neutrons might continue to split other atoms, producing a continuous process of chain reaction. Thus in 1939 (and possibly in 1934 in Italy) the atom had been split, but not in America.

Refugee scientists fleeing Hitler's Germany and Stalin's Russia soon brought the revolution in physics to America. In early 1939 the Hungarian physicist Leo Szilard, a reader of H. G. Wells's prophetic book, learned of the Hahn-Strassmann fission experiment and became concerned. Szilard, then a refugee living in New York, correctly anticipated that a new world of atomic weapons would be "headed for grief," especially if such weapons came into the hands of Adolf Hitler's Nazi Germany.

In 1939, with the support of influential patrons such as the banker Lewis Strauss, Szilard and another young Hungarian refugee, Edward Teller, persuaded the world-famous refugee physicist Einstein, living in Princeton, New Jersey, to sign a letter to President Franklin Delano Roosevelt warning him of the dangers of nuclear weapons and of the possibility that Hitler would soon acquire the valuable Joachimsthal uranium mines in Czechoslovakia. The letter was conveyed to Roosevelt shortly after World War II began with the German Blitzkrieg against Poland on September 1, 1939.

Despite a proliferation of committees on "the uranium question," little was accomplished in the United States until after Pearl Harbor. The initial key theoretical work on the possibility of building an atomic bomb was performed in England in 1940 and 1941 at Birmingham by two refugee German physicists, Rudolf Peierls and Otto Frisch. In the spring of 1940 they produced a working paper for the British government, then fighting for its life against aerial attacks by the Luftwaffe, on the theory of an atomic bomb. In it they concluded that a relatively small amount of uranium could constitute a "critical mass," which, if properly brought together at sufficient speed, could yield an explosion equivalent to many thousands of tons of TNT. The conclusions of Frisch and Peierls were reinforced by a second study carried out in 1941 by a secret British government working group known as the "MAUD Committee."

For further information on the early history of nuclear energy, see Ronald Clark, The Greatest Power on Earth: The International Race for Nuclear Supremacy from Earliest Theory to Three Mile Island (New York: Harper & Row, 1980). The papers and correspondence of Leo Szilard have been collected by his widow, Gertrude Weiss Szilard, and Spencer R. Weart as

Leo Szilard: His Version of the Facts; Selected Recollections and Correspondence, *vol. 2 (Cambridge, Mass., and London: MIT Press, 1980). The British story is told best in Margaret Gowing, Britain and Atomic Energy 1939–1945 (London: Macmillan, 1964).*

1. Nuclear Energy: H. G. Wells's Vision, 1914

It was in 1953 that the first Holsten-Roberts engine brought induced radio-activity into the sphere of industrial production, and its first general use was to replace the steam-engine in electrical generating stations. Hard upon the appearance of this came the Dass-Tata engine—the invention of two among the brilliant galaxy of Bengali inventors the modernisation of Indian thought was producing at this time—which was used chiefly for automobiles, aeroplanes, water-planes and such-like mobile purposes. The American Kemp engine, differing widely in principle but equally practicable, and the Krupp-Erlanger came hard upon the heels of this, and by the autumn of 1954 a gigantic replacement of industrial methods and machinery was in progress all about the habitable globe. Small wonder was this when the cost even of these earliest and clumsiest of atomic engines is compared with that of the power they superseded. Allowing for lubrication, the Dass-Tata engine, once it was started, cost a penny to run thirty-seven miles, and added only nine and a quarter pounds to the weight of the carriage it drove. It made the heavy alcohol-driven automobile of the time ridiculous in appearance as well as preposterously costly. For many years the price of coal and every form of liquid fuel had been clambering to levels that made even the revival of the draft-horse seem a practicable possibility, and now, with the abrupt relaxation of this stringency, the change in appearance of the traffic upon the world's roads was instantaneous. In three years the frightful armoured monsters that had hooted and smoked and thundered about the world for four awful decades were swept away to the dealers in old metal, and the highways thronged with light and clean and shimmering shapes of silvered steel. At the same time a new impetus was given to aviation by the relatively enormous power for weight

This selection first appeared in H. G. Wells, *The World Set Free* (New York: Dutton, 1914), pp. 51–53, 114–19, 152–53. Used with permission of the Executors of the Estate of H. G. Wells.

of the atomic engine; it was at last possible to add Redmayne's ingenious helicopter ascent and descent engine to the vertical propeller that had hitherto been the sole driving force of the aeroplane without overweighing the machine, and men found themselves possessed of an instrument of flight that could hover or ascend or descend vertically and gently as well as rush wildly through the air. The last dread of flying vanished. As the journalists of the time phrased it, this was the epoch of the Leap into Air. The new atomic aeroplane became indeed a mania; everyone of means was frantic to possess a thing so controllable, so secure, and so free from the dust and danger of the rod, and in France alone in the year 1943 thirty thousand of these new aeroplanes were manufactured and licensed and soared humming softly into the sky.

Never before in the history of warfare had there been a continuing explosive; indeed, up to the middle of the twentieth century the only explosives known were combustibles whose explosiveness was due entirely to their instantaneousness; and these atomic bombs which science burst upon the world that night were strange even to the men who used them. Those used by the Allies were lumps of pure Carolinum, painted on the outside with unoxidised cydonator inducive enclosed hermetically in a case of membranium. A little celluloid stud between the handles by which the bomb was lifted was arranged so as to be easily torn off and admit air to the inducive, which at once became active and set up radio-activity in the outer layer of the Carolinum sphere. This liberated fresh inducive, and so in a few minutes the whole bomb was blazing continual explosion. The Central European bombs were the same, except that they were larger and had a more complicated arrangement for animating the inducive.

Always before in the development of warfare the shells and rockets fired had been but momentarily explosive, they had gone off in an instant once and for all, and if there was nothing living or valuable within reach of the concussion and the flying fragments, then they were spent and over. But Carolinum, which belonged to the β-Group of Hyslop's so-called "suspended degenerator" elements, once its degenerative process had been induced, continued a furious radiation of energy, and nothing could arrest it. Of all Hyslop's artificial elements, Carolinum was the most heavily stored with energy and the most dangerous to make and handle. To this day it remains the most potent degenerator known. What the earlier twentieth-century chemists called its half period was seventeen days; that is to say, it poured out half of the huge store of energy in its great molecules in the space of seventeen days, the next seventeen days' emission was a half of that first period's outpouring, and so on. As with all radioactive substances, this Carolinum, though every seventeen days its power

is halved, though is never entirely exhausted, and to this day the battle-
fields and bomb-fields of that frantic time in human history are sprinkled
with radiant matter and so centres of inconvenient rays. . . .

What happened then when the celluloid stud was opened was that the
inducive oxydised and became active. Then the surface of the Carolinum
began to degenerate. This degeneration passed only slowly into the substance of the bomb. A moment or so after its explosion began it was still
mainly an inert sphere exploding superficially, a big, inanimate nucleus
wrapped in flame and thunder. Those that were thrown from aeroplanes
fell in this state; they reached the ground still mainly solid and, melting
soil and rock in their progress, bored into the earth. There, as more and
more of the Carolinum became active, the bomb spread itself out into a
monstrous cavern of fiery energy at the base of what became very speedily
a miniature active volcano. The Carolinum, unable to disperse freely,
drove into and mixed up with a boiling confusion of molten soil and super-
heated steam, and so remained, spinning furiously and maintaining an
eruption that lasted for years or months or weeks according to the size of
the bomb employed and the chances of its dispersal. Once launched, the
bomb was absolutely unapproachable and uncontrollable until its forces
were nearly exhausted, and from the crater that burst open above it, puffs
of heavy incandescent vapour and fragments of viciously punitive rock and
mud, saturated with Carolinum, and each a centre of scorching and blister-
ing energy, were flung high and far.

Such was the crowning triumph of military science, the ultimate explosive, that was to give the "decisive touch" to war. . . .

A recent historical writer has described the world of that time as one
that "believed in established words and was invincibly blind to the obvious
in things." Certainly it seems now that nothing could have been more
obvious to the people of the early twentieth century than the rapidity with
which war was becoming impossible. And as certainly they did not see it.
They did not see it until the atomic bombs burst in their fumbling hands.
Yet the broad facts must have glared upon any intelligent mind. All
through the nineteenth and twentieth centuries the amount of energy that
men were able to command was continually increasing. Applied to warfare
that meant that the power to inflict a blow, the power to destroy, was
continually increasing. There was no increase whatever in the ability to
escape. Every sort of passive defence, armour, fortifications, and so forth,
was being outmastered by this tremendous increase on the destructive
side. Destruction was becoming so facile that any little body of malcontents
could use it; it was revolutionising the problems of police and internal rule.
Before the last war began it was a matter of common knowledge that a man

could carry about in a handbag an amount of latent energy sufficient to wreck half a city. These facts were before the minds of everybody; the children in the streets knew them. And yet the world still, as the Americans used to phrase it, "fooled around" with the paraphernalia and pretensions of war.

It is only by realising this profound, this fantastic divorce between the scientific and intellectual movement on the one hand and the world of the lawyer-politician on the other that the men of a later time can hope to understand this preposterous state of affairs. Social organisation was still in the barbaric stage. There were already great numbers of actively intelligent men and much private and commercial civilisation, but the community as a whole was aimless, untrained, and unorganised to the pitch of imbecility. Collective civilisation, the "Modern State," was still in the womb of the future. . . .

For the whole world was flaring then into a monstrous phase of destruction. Power after power about the armed globe sought to anticipate attack by aggression. They went to war in a delirium of panic, in order to use their bombs first. China and Japan had assailed Russia and destroyed Moscow, the United States had attacked Japan, India was in anarchistic revolt with Delhi a pit of fire spouting death and flame; the redoubtable King of the Balkans was mobilising. It must have seemed plain at last to everyone in those days that the world was slipping headlong to anarchy. By the spring of 1959 from nearly two hundred centres, and every week added to their number, roared the unquenchable crimson conflagrations of the atomic bombs, the flimsy fabric of the world's credit had vanished, industry was completely disorganised, and every city, every thickly populated area was starving or trembled on the verge of starvation. Most of the capital cities of the world were burning; millions of people had already perished, and over great areas government was at an end. Humanity has been compared by one contemporary writer to a sleeper who handles matches in his sleep and wakes to find himself in flames.

For many months it was an open question whether there was to be found throughout all the race the will and intelligence to face these new conditions and make even an attempt to arrest the downfall of the social order. For a time the war spirit defeated every effort to rally the forces of preservation and construction. Leblanc seemed to be protesting against earthquakes and as likely to find a spirit of reason in the crater of Etna. Even though the shattered official Governments now clamored for peace, bands of irreconcilables and invincible patriots, usurpers, adventurers and political desperadoes were everywhere in possession of the simple apparatus for the disengagement of atomic energy and the initiation of new centres of destruction. The stuff exercised an irresistible fascination upon a certain type of mind. Why should anyone give in while he can still destroy his enemies? Surrender? While there is still a chance of blowing them to dust? The power of destruction which had once been the ultimate privilege of government was now the only power left in the world—and it was everywhere.

2. Leo Szilard and the Discovery of Fission

A. Szilard to Sir Hugo Hirst

6, Halliwick Road
London, N.10.
17th March, 1934

Dear Sir Hugo,
    As you are on holiday you might find pleasure in reading a few pages out of a book by H. G. Wells which I am sending you. I am certain you will find the first three paragraphs of Chapter The First (The New Source of Energy, page 42) interesting and amusing, whereas the other parts of the book are rather boring. It is remarkable that Wells should have written those pages in 1914.
    Of course, all this is moonshine, but I have reason to believe that in so far as the industrial applications of the present discoveries in physics are concerned, the forecast of the writers may prove to be more accurate than the forecast of the scientists. The physicists have conclusive arguments as to why we cannot create at present new sources of energy for industrial purposes; I am not so sure whether they do not miss the point.
    It is perhaps possible to be more definite some time after your return, and in the meantime I hope you will in any case enjoy glancing through these few pages.
    With best wishes for a pleasant stay,

Yours very truly,
Leo Szilard

These letters appeared in S. R. Weart and Gertrud Weiss, eds., *Leo Szilard: His Version of the Facts*, vol. 2 (Cambridge, Mass.: MIT Press, 1978), pp. 38–40, 62–65.

B. (addressee unknown)

28th July, 1934

Memorandum of Possible Industrial Applications Arising out of a New Branch of Physics

It is possible to indicate methods which might be successfully applied for the purpose of liberating atomic energy. It is not possible to foretell with certainty that these methods will be successful, but the experiments necessary for ascertaining this are fairly simple and could be carried out on a small scale in the university laboratories. Should such experiments give favourable results, the production of energy and its use for power production would be possible on such a large scale and probably with so little cost that a sort of industrial revolution could be expected; it appears doubtful, for instance, whether coal mining or oil production could survive after a couple of years.

I have applied for a group of patents in order to obtain patent protection for those methods which seemed to me promising, and it appears that these patents were successful in foreshadowing the latest developments in physics.

They include, for instance, methods for the artificial production of radio-active bodies based on a process which recently has been discovered by Fermi. The production of artificial "radium" for medical purposes based on these processes seems to be a sound commercial proposition, but it would be sidetracking the issue to concentrate on this point.

Facilities are required for two different purposes:

1. In order to develop and maintain a group of valid patents £500 are required for the next year, which would also take care of administrative expenses connected with the maintenance of the patents.

2. If we wish to start the necessary experiments one ought to secure the continuity of work for two to three years. It is not possible to state exactly what facilities will be required as this will depend to a large extent on what facilities will be provided by the university laboratory which would be used as a frame for this work. It would, however, be advisable to have £2,000 available for expenditures that may be incurred.

From the point of view of a financier who could consider contributing to the required facilities the position is this: the chances that the envisaged experiments will yield a favourable result may be estimated to anything between 1 to 20 and 1 to 5. The value of the return in case of success is, of course, enormous and could hardly be estimated in terms of money, so that from the purely financial point of view it is a sort of lottery with a fairly good chance to win a prize and enormous prizes. Yet it would be highly preferable to get financial support from quarters that would consider the experiment as a research work in the field of science which has a good chance of highly significant industrial applications, and realise that the exploitation of discoveries of this scope must not be organised on a purely commercial basis.

Difficulty will undoubtedly arise from the fact that it is not easy for anybody to form an independent opinion of some of the professors of the University of London who are working themselves in this field, and with whom I can easily keep in touch on the matter.

C. Letters from Szilard to Lewis Strauss

Hotel King's Crown
Opposite Columbia University
420 West 116th Street
New York City
January 25th, 1939

Mr. Lewis L. Strauss
% Kuhn, Loeb & Co.
52 William Street
New York City

Dear Mr. Strauss:

I feel that I ought to let you know of a very sensational new development in nuclear physics. In a paper in the *Naturwissenschaften* Hahn reports that he finds when bombarding uranium with neutrons the uranium breaking up into two halves giving elements of about half the atomic weight of uranium. This is entirely unexpected and exciting news for the average physicist. The Department of Physics at Princeton, where I spent the last few days, was like a stirred-up ant heap.

Apart from the purely scientific interest there may be another aspect of this discovery, which so far does not seem to have caught the attention of those to whom I spoke. First of all it is obvious that the energy released in this new reaction must be very much higher than in all previously known cases. It may be 200 million (electron-) volts instead of the usual 3–10 million volts. This in itself might make it possible to produce power by means of nuclear energy, but I do not think that this possibility is very exciting, for if the energy output is only two or three times the energy input, the cost of investment would probably be too high to make the process worthwhile. Unfortunately, most of the energy is released in the form of heat and not in the form of radioactivity.

I see, however, in connection with this new discovery potential possibilities in another direction. These might lead to a large-scale production of energy and radioactive elements, unfortunately also perhaps to atomic bombs. This new discovery revives all the hopes and fears in this respect which I had in 1934 and 1935, and which I have as good as abandoned in the course of the last two years. At present I am running a high temperature and am therefore confined to my four walls, but perhaps I can tell you more about these new developments some other time. Meanwhile you may look out for a paper in "Nature" by Frisch and Meitner which will soon appear and which might give you some information about this new discovery.

With best wishes,
Yours sincerely,
Leo Szilard

Hotel King's Crown
New York City
February 13th, 1939

Mr. Lewis L. Strauss
Brandy Rock Farm
Brandy, Va.

Dear Mr. Strauss:

I hope you and Mrs. Strauss enjoyed staying at Palm Beach and that you are now having a nice time at your farm.

After I left your train in Washington I spent a day with Dr. Teller there and another day with Dr. Wigner at Princeton and told both of them of our tentative plan to make use of the form of an "association" and let such an association take action if it seems desirable that something should be done along the lines which we discussed. Dr. Teller, who is Professor for Theoretical Chemistry at George Washington University, will be at our disposal if it becomes necessary to keep some person close to the Administration informed of the developments, and he also can get the cooperation of his colleagues in Washington if this will be required. Dr. Wigner thought that some of the experiments which we discussed could be done at Princeton. As he is an old friend of mine and has much influence in the department there is very much in favor of following his suggestions, but I feel that it will be necessary to see that the position is from the point of view of equipment, and whether some younger members of the department could cooperate without abandoning work in which they are at present engaged.

On my return to New York I went to see Fermi to tell him of all these conversations and also to discuss some of the small scale experiments which might be made in the near future.

Since my return almost every day some new information about uranium became available, and whenever I decided to do something one day it appeared foolish in the light of the new information on the next day. I found that the Radium Chemical Co. had in stock 200 mgm of radium mixed with beryllium, which is a nice constant source of fast neutrons. The rent for six months amounts to $500.00. As Fermi thought that he would like to use such a neutron source for his experiment I felt that I ought to get it for him. I did not see fair to ask you to take any decisions from a distance, and so I thought it might be best that I should advance $500.00 for expenses of this type and to see later whether you could sanction the expenditures afterwards. A few days later it turned out that this neutron source was too bulky to be suitable for Fermi's experiment, and Fermi said that for the present he is quite satisfied with the radon sources which he is getting anyway once a week at Columbia. In these circumstances I arranged with the Radium Chemical Co. that they will let me have one gram of radium on loan instead. This radium used in conjunction with the beryllium block sent from Oxford represents an intense source of photo-neutrons which can be used for a number of experiments. The rent is $125.00 per month, and we have to rent it for a minimum period of three months.

The outlook has changed in some important respects since I last saw you. It is now known that fast neutrons split both uranium and thorium, but slow neutrons do not split thorium, they probably do not split the bulk of uranium either. If enough neutrons are emitted when fast neutrons split thorium or uranium it will still be necessary to see whether or not the emitted neutrons are slowed down to a velocity at which they are ineffective before they had a chance to split enough nuclei to make the mainte-nance of a chain reaction possible.

On the other hand, slow neutrons seem to split a uranium isotope which is present in an abundance of about 1% in uranium. If this isotope could be used for maintaining chain reactions, it would have to be separated from the bulk of uranium. This, no doubt, would be done if necessary, but it might take five or ten years before it can be done on a technical scale. Should small scale experiments show that the uranium would, we would have the task immediately to attack the question of concentrating the rare isotope of uranium.

As you can see, the number of possibilities had increased since you left town. Some of the experiments which were devised, in particular the experiment which Fermi first planned, appear now to be much more difficult than before. Other experiments, such as those with photo-neutrons, are not affected, but of course they have somewhat the character of preliminary experiments.

I am enclosing a clipping which might interest you, as it shows the state of mind of some physicists on February 4th. The man who inspired this article did his best to hide what he thought, but his dementi is somewhat clumsy, and he almost gives himself away in the last paragraph.

Anyway, things have calmed down to some extent in the last few days, and the newspapers at least might soon forget about uranium.

With best wishes,

Yours very sincerely,
Leo Szilard

3. The Szilard-Einstein Letter, August 2, 1939

Albert Einstein
Old Grove Road
Peconic, Long Island
August 2nd, 1939

F. D. Roosevelt
President of the United States
White House
Washington, D.C.

Sir:

Some recent work by E. Fermi and L. Szilard, which has been communicated to me in manuscript, leads me to expect that the element uranium may be turned into a new and important source of energy in the immediate future. Certain aspects of the situation which has arisen seem to call for watchfulness and, if necessary, quick action on the part of the Administration. I believe therefore that it is my duty to bring to your attention the following facts and recommendation.

In the course of the last four months it has been made probable through the work of Joliot in France as well as Fermi and Szilard in America—that it may become possible to set up a nuclear chain reaction in a large mass of uranium, by which vast amounts of power and large quantities of new radium-like elements would be generated. Now it appears almost certain that this could be achieved in the immediate future.

This new phenomenon would also lead to the construction of bombs, and it is conceivable—though much less certain—that extremely powerful bombs of a new type may thus be constructed. A single bomb of this type, carried by boat and exploded in a port, might very well destroy the whole port together with some of the surrounding territory. However, such bombs might very well prove to be too heavy for transportation by air.

The United States has only very poor ores of uranium in moderate quantities. There is some good ore in Canada and the former Czechoslovakia, while the most important source of uranium is Belgian Congo.

In view of this situation you may think it desirable to have some permanent contact maintained between the Administration and the group of physicists working on chain reactions in America. One possible way of achieving this might be for you to entrust with this task a person who has your confidence and who could perhaps serve in an unofficial capacity. His task might comprise the following:

a) to approach Government Departments, keep them informed of the further development, and put forward recommendations for Government action, giving particular attention to the problem of securing a supply of uranium ore for the United States.

b) to speed up the experimental work, which is at present being carried on within the limits of the budgets of University laboratories, by providing funds, if such funds be required, through his contacts with private persons who are willing to make contributions for this cause, and perhaps also by obtaining the co-operation of industrial laboratories which have the necessary equipment.

I understand that Germany has actually stopped the sale of uranium from the Czechoslovakian mines which she has taken over. That she should have taken such early action might perhaps be understood on the ground that the son of the German Under-Secretary of State, von Weizsacker, is attached to the Kaiser-Wilhelm-Institut in Berlin where some of the American work on uranium is now being repeated.

Yours very truly,
/ s / Albert Einstein

4. The Frisch-Peierls Memorandum, 1940

On the Construction of a "Super-bomb"; based on a
Nuclear Chain Reaction in Uranium

The possible construction of "super-bombs" based on a nuclear chain reaction in uranium has been discussed a great deal and arguments have been brought forward which seemed to exclude this possibility. We wish here to point out and discuss a possibility which seems to have been over-looked in these earlier discussions.

Uranium consists essentially of two isotopes, ^{238}U (99.3%) and ^{235}U (0.7%). If a uranium nucleus is hit by a neutron, three processes are possible: (1) scattering, whereby the neutron changes direction and, if its energy is above about 0.1 MeV, loses energy; (2) capture, when the neutron is taken up by the nucleus; and (3) fission, i.e. the nucleus breaks up into two nuclei of comparable size, with the liberation of an energy of about 200 MeV.

The possibility of a chain reaction is given by the fact that neutrons are emitted in the fission and that the number of these neutrons per fission is greater than 1. The most probable value for this figure seems to be 2.3, from two independent determinations.

However, it has been shown that even in a large block of ordinary uranium no chain reaction would take place since too many neutrons would

This selection appeared in Margaret Gowing, *Britain and Atomic Energy 1939–1945* (New York: MacMillan, 1964), pp. 389–93.

be slowed down by inelastic scattering into the energy region where they are strongly absorbed by ^{238}U.

Several people have tried to make chain reactions possible by mixing the uranium with water, which reduces the energy of the neutrons still further and thereby increases their efficiency again. It seems fairly certain, however that even then it is impossible to sustain a chain reaction.

In any case, no arrangement containing hydrogen and based on the action of slow neutrons could act as an effective super-bomb, because the reaction would be too slow. The time required to slow down a neutron is about 10^{-5} sec and the average time loss before a neutron hits a uranium nucleus is even 10^{-4} sec. In the reaction, the number of neutrons would increase exponentially, like e^{t}/τ where τ would be at least 10^{-4} sec. When the temperature reaches several thousand degrees the container of the bomb will break and within 10^{-4} sec the uranium would have expanded sufficiently to let the neutrons escape and so to stop the reaction. The energy liberated would, therefore, be only a few times the energy required to break the container, i.e. of the same order of magnitude as with ordinary high explosives.

Bohr has put forward strong arguments for the suggestion that the fission observed with slow neutrons is to be ascribed to the rare isotope ^{235}U, and that this isotope has, on the whole, a much greater fission probability than the common isotope ^{238}U. Effective methods for the separation of isotopes have been developed recently, of which the method of thermal diffusion is simple enough to permit separation on a fairly large scale.

This permits, in principle, the use of nearly pure ^{235}U in such a bomb, a possibility which apparently has not so far been seriously considered. We have discussed this possibility and come to the conclusion that a moderate amount of ^{235}U would indeed constitute an extremely efficient explosive.

The behaviour of ^{235}U under bombardment with fast neutrons is not known experimentally, but from rather simple theoretical arguments it can be concluded that almost every collision produces fission and that neutrons of any energy are effective. Therefore it is not necessary to add hydrogen, and the reaction, depending on the action of fast neutrons, develops with very great rapidity so that a considerable part of the total energy is liberated before the reaction gets stopped on account of the expansion of the material.

The critical radius γ_0—i.e. the radius of a sphere in which the surplus of neutrons created by the fission is just equal to the loss of neutrons by escape through the surface—is, for a material with a given composition, in a fixed ratio to the mean free path of neutrons, and this in turn is inversely proportional to the density. It therefore pays to bring the material into the densest possible form, i.e. the metallic state, probably sintered or hammered. If we assume for ^{235}U, no appreciable scattering, and 2.3 neutrons emitted per fission, then the critical radius is found to be 0.8 times the mean free path. In the metallic state (density 15), and assuming a fission cross-section of 10^{-23} cm^2, the mean free path would be 2.6 cm and γ_0 would be 2.1 cm, corresponding to a mass of 600 grams. A sphere of metallic ^{235}U of a radius greater than γ_0 would be explosive, and one might think of about 1 kg as a suitable size for the bomb.

The speed of the reaction is easy to estimate. The neutrons emitted in the fission have velocities of about 10^{-9} cm/sec and they have to travel 2.6 cm before hitting a uranium nucleus. For a sphere well above the critical size the loss through neutron escape would be small, so we may assume that each neutron after a life of 2.6×10^{-9} sec, produces fission, giving birth to two neutrons. In the expression e^t/τ for the increase of neutron density with time, it would be about 4×10^{-9} sec, very much shorter than in the case of a chain reaction depending on slow neutrons.

If the reaction proceeds until most of the uranium is used up, temperatures of the order of 10^{10} degrees and pressure of about 10^{13} atmospheres are produced. It is difficult to predict accurately the behaviour of matter under these extreme conditions, and the mathematical difficulties of the problem are considerable. By a rough calculation we get the following expression for the energy liberated before the mass expands so much that the reaction is interrupted:

$E = 0.2M(\gamma^2/\tau^2)(\sqrt{(\gamma/\gamma_0)}-1)$

(M, total mass of uranium; γ, radius of sphere; γ_0, critical radius; τ, time required for neutron density to multiply by a factor e). For a sphere of radius 4.2 cm ($\gamma_0 = 2.1$ cm), M = 4700 grams, $\tau = 4 \times 10^{-9}$ sec, we find E = 4×10^{20} ergs, which is about one-tenth of the total fission energy. For a radius of about 8 cm (M = 32 kg) the whole fission energy is liberated, according to formula (1). For small radii the efficiency falls off even faster than indicated by formula (1) because τ goes up as γ approaches γ_0. The energy liberated by a 5 kg bomb would be equivalent to that of several thousand tons of dynamite, while that of a 1 kg bomb, though about 500 times less, would still be formidable.

It is necessary that such a sphere should be made in two (or more) parts which are brought together first when the explosion is wanted. Once assembled, the bomb would explode within a second or less, since one neutron is sufficient to start the reaction and there are several neutrons passing through the bomb in every second, from the cosmic radiation. (Neutrons originating from the action of uranium alpha rays on light-element impurities would be negligible provided the uranium is reasonably pure.) A sphere with a radius of less than about 3 cm could be made up in two hemispheres, which are pulled together by springs and kept separated by a suitable structure which is removed at the desired moment. A larger sphere would have to be composed of more than two parts, if the parts, taken separately, are to be stable.

It is important that the assembling of the parts should be done as rapidly as possible, in order to minimise the chance of a reaction getting started at a moment when the critical conditions have only just been reached. If this happened, the reaction rate would be much slower and the energy liberation would be considerably reduced; it would, however, always be sufficient to destroy the bomb.

It may be well to emphasise that a sphere only slightly below the critical size is entirely safe and harmless. By experimenting with spheres of gradually increasing size and measuring the number of neutrons emerging from them under a known neutron bombardment, one could accurately determine the critical size, without any danger of a premature explosion.

For the separation of the ^{235}U, the method of thermal diffusion, developed by Clusius and others, seems to be the only one which can cope with the large amounts required. A gaseous uranium compound, for example uranium hexafluoride, is placed between two vertical surfaces which are kept at a different temperature. The light isotope tends to get more concentrated near the hot surface, where it is carried upwards by the convection current. Exchange with the current moving downwards along the cold surface produces a fractionating effect, and after some time a state of equilibrium is reached when the gas near the upper end contains markedly more of the light isotope than near the lower end.

For example, a system of two concentric tubes, of 2 mm separation and 3 cm diameter, 150 cm long, would produce a difference of about 40% in the concentration of the rare isotope between its ends, and about 1 gram per day could be drawn from the upper end without unduly upsetting the equilibrium.

In order to produce large amounts of highly concentrated ^{235}U, a great number of these separating units will have to be used, being arranged in parallel as well as in series. For a daily production of 100 grams of ^{235}U of 90% purity, we estimate that about 100,000 of these tubes would be required. This seems a large number, but it would undoubtedly be possible to design some kind of a system which would have the same effective area in a more compact and less expensive form.

In addition to the destructive effect of the explosion itself, the whole material of the bomb would be transformed into a highly radioactive stage. The energy radiated by these active substances will amount to about 20% of the energy liberated in the explosion, and the radiations would be fatal to living beings even a long time after the explosion.

The fission of uranium results in the formation of a great number of active bodies with periods between, roughly speaking, a second and a year. The resulting radiation is found to decay in such a way that the intensity is about inversely proportional to the time. Even one day after the explosion the radiation will correspond to a power expenditure of the order of 1,000 kW, or to the radiation of a hundred tons of radium.

Any estimates of the effects of this radiation on human beings must be rather uncertain because it is difficult to tell what will happen to the radioactive material after the explosion. Most of it will probably be blown into the air and carried away by the wind. This cloud of radioactive material will kill everybody within a strip estimated to be several miles long. If it rained the danger would be even worse because active material would be carried down to the ground and stick to it, and persons entering the contaminated area would be subjected to dangerous radiations even after days. If 1% of the active material sticks to the debris in the vicinity of the explosion and if the debris is spread over an area of, say, a square mile, any person entering this area would be in serious danger, even several days after the explosion.

In the estimates, the lethal dose of penetrating radiation was assumed to be 1,000 Roentgen; consultation of a medical specialist on X-ray treatment and perhaps further biological research may enable one to fix the danger limit more accurately. The main source of uncertainty is our lack of knowledge as to the behaviour of materials in such a super-explosion, and an expert on high explosives may be able to clarify some of these problems.

Effective protection is hardly possible. Houses would offer protection only at the margins of the danger zone. Deep cellars or tunnels may be comparatively safe from the effects of radiation, provided air can be supplied from an uncontaminated area (some of the active substances would be noble gases which are not stopped by ordinary filters).

The irradiation is not felt until hours later when it may be too late. Therefore it would be very important to have an organisation which determines the exact extent of the danger area, by means of ionisation measurements, so that people can be warned from entering it.

O. R. Frisch
R. Peierls

(The University, Birmingham)

5. The MAUD Report, 1941

Report by M.A.U.D. Committee On the Use of
Uranium for a Bomb

PART I

1. General Statement

Work to investigate the possibilities of utilising the atomic energy of uranium for military purposes has been in progress since 1939, and a stage has now been reached when it seems desirable to report progress.

We should like to emphasise at the beginning of this report that we entered the project with more scepticism than belief, though we felt it was a matter which had to be investigated. As we proceeded we became more and more convinced that release of atomic energy on a large scale is possible and that conditions can be chosen which would make it a very powerful weapon of war. We have now reached the conclusion that it will be possible to make an effective uranium bomb which, containing some 25 lb of active material, would be equivalent as regards destructive effect to 1,800 tons of T.N.T. and would also release large quantities of radioactive substance, which would make places near to where the bomb exploded dangerous to human life for a long period. The bomb would be composed of an active constituent (referred to in what follows as ^{235}U) present to the extent of about a part in 140 in ordinary Uranium. Owing to the very small difference in properties (other than explosive) between this substance and the rest of the Uranium, its extraction is a matter of great difficulty and a plant to produce 2¼ lb (1 kg) per day (or 3 bombs per month) is estimated to cost approximately £5,000,000 pounds, of which sum a considerable proportion would be spent on engineering, requiring labour of the same highly skilled character as is needed for making turbines.

In spite of this very large expenditure we consider that the destructive effect, both material and moral, is so great that every effort should be made to produce bombs of this kind. As regards the time required, Imperial Chemical Industries after consultation with Dr. Guy of Metropolitan-Vickers, estimate that the material for the first bomb could be ready by the

This selection appeared in Margaret Gowing, *Britain and Atomic Energy* 1939–1945 (New York: MacMillan, 1964), pp. 394–98.

end of 1943. This of course assumes that no major difficulty of an entirely unforeseen character arises. Dr. Ferguson of Woolwich estimates that the time required to work out the method of producing high velocities required for fusing (see paragraph 3) is 1–2 months. As this could be done concurrently with the production of the material no further delay is to be anticipated on this score. Even if the war should end before the bombs are ready the effort would not be wasted, except in the unlikely event of complete disarmament, since no nation would care to risk being caught without a weapon of such decisive possibilities.

We know that Germany has taken a great deal of trouble to secure supplies of the substance known as heavy water. In the earlier stages we thought that this substance might be of great importance for our work. It appears in fact that its usefulness in the release of atomic energy is limited to processes which are not likely to be of immediate war value, but the Germans may by now have realised this, and it may be mentioned that the lines on which we are now working are such as would be likely to suggest themselves to any capable physicist.

By far the largest supplies of Uranium are in Canada and the Belgian Congo, and since it has been actively looked for because of the radium which accompanies it, it is unlikely that any considerable quantities exist which are unknown except possibly in unexplored regions.

2. Principle Involved

This type of bomb is possible because of the enormous store of energy resident in atoms and because of the special properties of the active constituent of uranium. The explosion is very different in its mechanism from the ordinary chemical explosion, for it can occur only if the quantity of ^{235}U is greater than a certain critical amount. Quantities of the material less than the critical amount are quite stable. Such quantities are therefore perfectly safe and this is a point which we wish to emphasise. On the other hand, if the amount of material exceeds the critical value it is unstable and a reaction will develop and multiply itself with enormous rapidity, resulting in an explosion of unprecedented violence. Thus all that is necessary to detonate the bomb is to bring together two pieces of the active material each less than the critical size but which when in contact form a mass exceeding it.

3. Method of Fusing

In order to achieve the greatest efficiency in an explosion of this type, it is necessary to bring the two halves together at high velocity and it is proposed to do this by firing them together with charges of ordinary explosive in a form of double gun.

The weight of this gun will of course greatly exceed the weight of the bomb itself, but should not be more than 1 ton, and it would certainly be within the carrying capacity of a modern bomber. It is suggested that the bomb (contained in the gun) should be dropped by parachute and the gun should be fired by means of a percussion device when it hits the ground. The time of drop can be made long enough to allow the aeroplane to escape from the danger zone, and as this is very large, great accuracy of aim is not required.

4. Probable Effect

The best estimate of the kind of damage likely to be produced by the explosion of 1,800 tons of T.N.T. is afforded by the great explosion at Halifax N.S. in 1917. The following account is from the *History of Explosives*. "The ship contained 450,000 lb of T.N.T., 122,960 lb of guncotton, and 4,661,794 lb of picric acid wet and dry, making a total of 5,234,754 lb. The zone of the explosion extended for about ¾ mile in every direction and in this zone the destruction was almost complete. Severe structural damage extended generally for a radius of 1-⅛ to 1-¼ miles, and in one direction up to 1-¾ miles from the origin. Missiles were projected to 3–4 miles, window glass broken up to 10 miles generally, and in one instance up to 61 miles."

In considering this description it is to be remembered that part of the explosives cargo was situated below water level and part above.

5. Preparation of Material and Cost

We have considered in great detail the possible methods of extracting the ^{235}U from ordinary uranium and have made a number of experiments. The scheme which we recommend is described in Part II of this report and in greater detail in Appendix IV. It involves essentially the gaseous diffusion of a compound of uranium through gauzes of very fine mesh.

In the estimates of size and cost which accompany this report, we have only assumed types of gauze which are at present in existence. It is probable that a comparatively small amount of development would enable gauzes of smaller mesh to be made and this would allow the construction of a somewhat smaller and consequently cheaper separation plant for the same output.

Although the cost per lb of this explosive is so great it compares very favourably with ordinary explosives when reckoned in terms of energy released and damage done. It is, in fact considerably cheaper, but the points which we regard as of overwhelming importance are the concentrated destruction which it would produce, the large moral effect, and the saving in air effort the use of this substance would allow, as compared with bombing with ordinary explosives.

6. Discussion

One outstanding difficulty of the scheme is that the main principle cannot be tested on a small scale. Even to produce a bomb of the minimum critical size would involve a great expenditure of time and money. We are however convinced that the principle is correct, and whilst there is still some uncertainty as to the critical size it is most unlikely that the best estimate we can make is so far in error as to invalidate the general conclusions. We feel that the present evidence is sufficient to justify the scheme being strongly pressed.

As regards the manufacture of the ^{235}U we have gone nearly as far as we can on a laboratory scale. The principle of the method is certain, and the application does not appear unduly difficult as a piece of chemical engineering. The need to work on a larger scale is now very apparent and we are beginning to have difficulty in finding the necessary scientific personnel. Further, if the weapon is to be available in say two years from now, it is necessary to start plans for the erection of a factory, though no really large expenditure will be needed till the 20-stage model has been tested. It is also important to begin training men who can ultimately act as supervisors of the manufacture. There are a number of auxiliary pieces of apparatus to be developed, such as those for measuring the concentration of the ^{235}U. In addition, work on a fairly large scale is needed to develop the chemical side for the production in bulk of uranium hexafluoride, the gaseous compound we propose to use.

It will be seen from the foregoing that a stage in the work has now been reached at which it is important that a decision should be made as to whether the work is to be continued on the increasing scale which would be necessary if we are to hope for it as an effective weapon for this war. Any considerable delay now would retard by an equivalent amount the date by which the weapon could come into effect.

7. Action in U.S.

We are informed that while the Americans are working on the uranium problem the bulk of their effort has been directed to the production of energy, as discussed in our report on uranium as a source of power, rather than to the production of a bomb. We are in fact co-operating with the United States to the extent of exchanging information, and they have undertaken one or two pieces of laboratory work for us. We feel that it is important and desirable that development work should proceed on both sides of the Atlantic irrespective of where it may be finally decided to locate the plant for separating the ^{235}U, and for this purpose it seems desirable that certain members of the committee should visit the United States. We are informed that such a visit would be welcomed by members of the United States committees which are dealing with this matter.

8. Conclusions and Recommendations

(i)   The committee considers that the scheme for a uranium bomb is practicable and likely to lead to decisive results in the war.

(ii) It recommends that this work be continued on the highest priority and on the increasing scale necessary to obtain the weapon in the shortest possible time.

(iii) That the present collaboration with America should be continued and extended especially in the region of experimental work.

II

The Manhattan Project

*B*y 1942 scientists in both England and the United States were convinced
that a nuclear weapon based on the principle of fission could be constructed
within a period of three to four years. Yet the complex process of separating
fissionable U235 or producing man-made plutonium, also fissionable, was
expensive and unproven.

On December 2, 1942, in a racquet court beneath the West Stands of
Stagg Field at the University of Chicago a team of scientists led by Nobel
Prize chemist Enrico Fermi succeeded in achieving the first controlled, self-
sustaining nuclear chain reaction, using a "pile" of graphite and uranium
blocks. Fermi promptly wired James Conant, project chief, that "the Italian
navigator has landed in the New World," signifying a successful experiment.
Within months ground was broken for three supersecret atomic cities: Hanford, Washington, where plutonium production would occur; Oak Ridge,
Tennessee, where uranium separation plants were under construction; and
Los Alamos, New Mexico, where bomb design, experimental testing, and
assembly would be centered. Thus the "Manhattan Engineer District" established under General Leslie Groves and the Army Corps of Engineers in
the summer of 1942 was not in Manhattan but in the scattered university
research laboratories and new atomic towns across the country.

In 1943 a team of scientists was formed at Los Alamos under the direction of J. Robert Oppenheimer, a brilliant physicist educated at Harvard
and Göttingen who had taught students at Berkeley and California Institute
of Technology in the 1930s. Oppenheimer was able to persuade General
Groves that the most auspicious approach to building an atomic bomb was
to have civilian scientists working in an isolated military environment at Los
Alamos, where free discussion could still occur. Groves wanted compart-
mentalization of all knowledge and information so that each individual would
know only what he or she needed to know. Oppie, as he was called, wanted

an open exchange of ideas. In the end Oppenheimer had his way, and Los Alamos became a legendary community of brilliant "crackpots," as Groves labelled them, working feverishly to create a nuclear weapon before Adolf Hitler might do the same.

In fact, Hitler had sacrificed his nuclear physicists to the caldron of the Eastern Front, where Stalin's Red Army held firm, and to the budgetary demands of German rocketry. Dependent on a single heavy water plant in Norway, and believing that graphite would not serve to moderate the flow of neutrons (thus making fission more probable), German scientists by 1944 had made no significant progress toward either a reactor or a bomb. By autumn 1944 German failure was known to American military and political leaders, and the focus shifted to Japan.

It was not until early 1945 that the first plutonium and uranium began to emerge from Hanford and Oak Ridge in usable quantities. In the meantime Los Alamos scientists had discovered that a gun-type uranium bomb was quite feasible, but that plutonium was so fissionable that it could predetonate unless a critical mass were achieved almost instantly. The new method, "imploding" a plutonium sphere by exploding chemical "lenses" around its surface, needed to be tested. The Hiroshima bomb needed no test; the Nagasaki bomb was first exploded in the sands of Alamogordo, New Mexico, in July 1945.

Since early 1943 American military leaders assumed that if and when an atomic bomb was developed it would be used first against Japan. By 1945 conventional mass firebombing had reduced most Japanese cities to rubble; the "Meetinghouse" raid over Tokyo in March killed more than 100,000 people. In this light it was not difficult for American military and scientific committees to decide to use the first nuclear weapons in order to bring the bloody fighting in the Pacific to an end.

When President Roosevelt died in April 1945, the new president, Harry S. Truman, had no knowledge of the Manhattan Project to build an atomic bomb at a cost approaching $2 billion. He had suddenly inherited a massive technological race against Germany, Japan, and the Soviet Union conducted almost entirely in secret. Truman delayed his trip to Potsdam long enough for the successful plutonium bomb test and then told Stalin in oblique terms about the existence of the new weapon. Only later did Truman learn that Stalin's noncommittal response veiled an intimate knowledge of the Manhattan Project gained through Soviet spies operating in the United States.

The decision to drop the atomic bomb has been and will continue to be the subject of a debate without end. Given the Japanese attack on Pearl Harbor on December 7, 1941, the time and money invested in the project, and the desire to end the war, the bureaucratic momentum toward dropping the bomb was enormous. Japanese surrender was not forthcoming, and an

American invasion estimated to cost half a million American lives was imminent. After some debate among scientists about the morality of using the bomb, as opposed to demonstrating its power to enemy observers, Truman made the decision to use it. On August 6, 1945, the Japanese city of Hiroshima was destroyed by an atomic bomb dropped from the "Enola Gay," a B-29 operating out of the island base of Tinian. On August 9 the city of Nagasaki, a secondary target, was likewise destroyed by a plutonium bomb. The Japanese surrendered within a few more days, and Truman was a hero for having made use of "the greatest thing in history" to bring a terrible war to an end.

The official history of the Manhattan Project is by Richard G. Hewlett and Oscar E. Anderson, Jr., The New World, 1939/1946, vol. 1 of A History of the United States Atomic Energy Commission (University Park: Pennsylvania State University Press, 1962). A valuable personal memoir is Arthur Holly Compton, Atomic Quest; A Personal Narrative (Oxford: Oxford University Press, 1956). The abridged version of the declassified multivolume project history is available in Anthony Cave Brown and Charles B. MacDonald, eds., The Secret History of the Atomic Bomb (New York: Delta, 1977). On the Los Alamos environment, see James W. Kunetka, City of Fire. Los Alamos and the Birth of the Atomic Age 1943–1945 (Englewood Cliffs, N.J.: Prentice-Hall, 1978). The best study of the diplomacy of the atomic bomb is Martin Sherwin, A World Destroyed. The Atomic Bomb and the Grand Alliance (New York: Random House, 1977). On Oppenheimer, see Peter Goodchild, J. Robert Oppenheimer; Shatterer of Worlds (Boston: Houghton Mifflin, 1981).

---

6. Letters of J. Robert Oppenheimer, 1942–45

A. To James B. Conant

Berkeley
November 30, 1942

Dear Dr. Conant:

Your letter reached me with some delay since I returned to Berkeley only a day or so ago. I should like to answer first your P.S. You are quite right that the purities listed in Groves' compulsory memo are a little mis-

These letters appeared in Alice Kimball Smith and Charles Weiner, eds., Robert Oppenheimer: Letters and Recollections (Cambridge, Mass.: Harvard University Press, 1980), pp. 240–46, 256, 262–63, 267–68, 270–72, 282, 286–87, 293, 294.

leading. The reason for this is that Groves defined a satisfactory bomb as one that had a 50 per cent chance of exceeding a 1,000-ton TNT equivalent. The absolute requirements are figured on this basis. It is, of course, my opinion that we should be wanton to strive for such a low goal, but I believe that some good was in fact done by indicating at that time that the purity requirements are not fantastic. The desirable requirements are equally undefined since the purer the material (up to a purity of about 100-fold that given) the less must we be worried about getting the maximum speed for the firing mechanism of the detonator; and this will make for simplicity and reliability in operation. In the Washington memo all impurities were listed on the assumption that not more than five elements would reach the tabulated values. I have, in the meantime, given a much more careful account of what the actual situation is to the committee. I met with them one day in Chicago, came out with them on the train, and have spent two days with them here in Berkeley, and we have had ample opportunity to discuss the purity question and many other aspects of our problem. The information which I have given them now is contained in a Chicago report on the feasibility of the 49 project and is as follows: If the concentrations by weight are as given in the accompanying table, then the chance of pre-detonation is 5 per cent, if only one element is present in the listed amount. If n elements are present in the listed amount the chance of pre-detonation is $5n$ per cent. The chance of a pre-detonation in which the energy release is less than 10,000 tons TNT equivalent is $0.5n$ per cent. In this range the effects of impurities are additive, and from the actual concentrations and figures listed one can figure out the probability of any given energy release. In any case, unless the firing mechanism fails completely the energy release will be more than sufficient to destroy the material and to make its recovery impossible. The figures given in the table are in part based on experimental values. In the case of O and C the figures represent highly conservative estimates based on the assumption that those isotopes which are dangerous will be as dangerous as the worst element, namely, Be.

Element Concentration by Weight

Be	10^{-7}	F	5×10^{-6}
Li	5×10^{-7}	Na	2×10^{-5}
B	2×10^{-7}	Mg	10^{-4}
C	2×10^{-5}	Al	2×10^{-5}
O	10^{-4}	Si	5×10^{-4}
		P	10^{-4}

(Some purity requirements on elements between P and Fe, none beyond Cu)

The only essential changes since the Washington memo are that we have sufficiently examined the experiments somewhat; and that we have studied the case of N carefully enough to be positive that there are no purity requirements on that element. The committee was of the opinion that the purity requirements as they now stand could, with a very high probability, be met. In fact, the Chicago uranium is good enough except for C and O, and they have made no effort at all to solve this problem. If it were necessary it would be possible to work with depleted C and O and so considerably relax the chemical conditions on these elements. In fact, the committee was of the opinion that the major extraction processes which have to be handled automatically and the removal of traces of active material, coupled with the necessity for working in lots of less than 100 grams or of introducing suitable neutron absorbers as "safers," would present greater technical difficulties than the purification. Nevertheless, in our last discussions they seemed convinced that the helium-cooled graphite pile was a good bet.

Now to the second point, the main subject of your letter, where I feel myself on less secure ground. It is, of course, natural that the men we are after will leave a big hole. I may though, in this connection, remind you that when McMillan himself left the Radiation Laboratory for San Diego there were the same dire predictions of disastrous disruption. Nevertheless, the Radiation Laboratory has not only survived but has, as you know, flourished and expanded. In view of this and of the very large number of men of the first rank who are now working on that project, I am inclined not to take too seriously the absolute no's with which we shall be greeted. I believe that it is important to emphasize that we should in any case be willing to let these men have time enough in their old positions to try to minimize the disruption of their leaving. I also agree that a fundamental clarification on this personnel problem, which can hardly be complete without Dr. Bush's participation, will be necessary. The job we have to do will not be possible without personnel substantially greater than that which we now have available, and I should only be misleading you and all others concerned with the S-1 project if I were to promise to get the work done without this help.

The suggestion of Eckhardt as a substitute for Kurie is a welcome one and we shall arrange to talk with him on our next trip east.[1] There are, however, two reasons more substantial than prejudice why the limitation to men who are known to us is sound: 1) that the technical details of this work will in large part have to do with atomic physics so that any man whose

1. Probably E. A. Eckhardt, a physicist who was vice-president of Gulf Research and Development Company in 1942.

experience has been in another field will necessarily be of more limited usefulness; Kurie, for instance, would have had as one of his responsibilities the installation and servicing of the cyclotron. The second reason is that in a tight isolated group such as we are now planning, some warmth and trust in personal relations is an indispensable prerequisite, and we are, of course, able to insure this only in the case of men whom we have known in the past. You will have had from me a note on possible alternatives to Kurie. If none of our suggestions seem practicable we shall see whether Dr. Eckhardt could fill the bill.

With good wishes,

Very sincerely yours,
Robert Oppenheimer

B. To Hans and Rose Bethe

(Berkeley, California)
December 28, 1942

Dear Hans and Rose:

Thank you for your fine letters and documents. I would like to answer as many of the questions that you raise as I now can so that you may be informed about the present state of affairs. Some of the points have not yet been settled and one or two of them have not been even raised. I shall keep your lists as a reminder of what we shall have to do.

First, the points of your letter:

1. We are trying now to put the laboratory, as distinct from the town, on the basis of an OSRD contract with a dummy contracting agent, say Cal Tech or Harvard. This will give us the flexibility in salary and procurement that we all so much desire. I am, however, assured that the Army procurement procedure has quite recently been radically altered. It is no longer necessary to obtain the conventional three bids; procurement can be carried out by our agent in Pasadena without any authorization and on the unique basis of getting what we need as fast as possible. It is our plan to pay all civilians $^{12}/_{10}$ of their university salary and for the younger men to pay on the M.I.T. salary scale. It is not yet clear what we shall do about subsistence. We may pay it and charge for rent, utilities, school, hospitals, etc. On the other hand we may not pay subsistence and provide housing, utilities and normal services in the town without charge. It seems to me that either of these schemes will be satisfactory. It is largely a question of

whether we wish to keep some form of control on the services (how much electricity and fuel is used, for instance) which procedure we shall follow.

2. We looked into the question of Dana P. Mitchell and were unable to get anywhere with it. At present there is a good possibility that Allison will come to us from Chicago; in fact, he visited the school with me on my last trip.[1] I also understand that Kurie is very anxious to come and as soon as we have given Conant and Bush a chance to live up to their commitments we will look into this. In the meantime, McMillan, Kennedy and I will try to get the rudiments of the procurement problem outlined.

3. I do not think that I can at this time refer Ramsey's name to Conant because of my promise, but one of the reasons that I am so optimistic about our future is that if we really get the men promised us there will be many others who will want to come and that should immensely simplify the job of getting them released.

Now to Rose's question. There will be a sort of city manager who will be distinct from Colonel Harmon.[2] Harmon was trying to find a man for this job and there is a man in New Mexico, a civilian, whom we could obtain if he is not successful. There will also be a city engineer and together they will take care of the problems outlined by you. We hope to persuade one of the teachers at the school to stay on to be our professional teacher. It is true that both Kay Manley and Elsie McMillan are professional school teachers and there will no doubt be others, but it seems to me unlikely that anyone with a very young child will be able to devote very much time to the community. There will be two hospitals, one in the town and one in the M.P. camp. The one in the town will be run by our doctor or doctors and she will be paid a regular salary so that she will be independent of our fees. I believe we shall have a group health plan of some kind but this is one of the questions that is tied up with how we handle subsistence.

Room is being provided for a laundry; each house will have its washtub; and we shall be able to send laundry to Santa Fe regularly. It may be necessary for us to provide the equipment for the group laundry since this is now frozen, but this is a point that is not yet settled.

We plan to have two eating places. There will be a regular mess for unmarried people which will be, when we are running at full capacity, just large enough to take care of these. The Army will take care of the help for

1. Dana P. Mitchell was a physicist at Columbia University, where he had charge of purchases and lab services. Samuel Allison, at this time director of the Met Lab, did not come to Los Alamos until 1944.

2. Colonel J. M. Harmon was the first commanding officer of Post Administration, which had charge of the construction and management of the town.

this and I do not know whether the personnel will be Army or civilian. We will also arrange to have a cafe where married people can eat out. This will probably be able to handle about twenty people at a time and will be a little fancy, and may be by appointment only. We are trying to persuade one of the natives to operate this and we have a good building for it.

There will be a recreation officer who will make it his business to see that such things as libraries, pack trips, movies, and so on are taken care of, and he will no doubt welcome the help of as many of us as are willing. The bachelor apartments will be run by the Army and will be completely served. The store will be a so-called Post Exchange which is a combination of country store and mail order house. That is, there will be stocks on hand and the Exchange will be able to order for us what they do not carry. There will be a vet to inspect the meat and barbers and such like. There will also be a cantina where we can have beer and cokes and light lunches.

The employment agency is something which we have not made any official provision for although we have discussed it with Harmon. I think it would be fine if you planned to help out with this as soon as you get there, and provision will be made to pay you for that work.

I am a little reluctant to do too much writing about the details of our life there until people are actually on the job, but I agree with you that some rudimentary information about the nature of the life and possibilities of work should be made available to everyone. I am answering as well as I can your questionnaire on the enclosed sheet.

That will have to be all for the moment. But I would appreciate it if you would get in touch with me if any new problems arise which we ought to consider, or if you think any of the arrangements that we have made so far are definitely unsatisfactory. Our best guarantee that they can be made satisfactory is in the great effort and generosity that Harmon and Groves have both brought to setting up this odd community and in their evident desire to make a real success of it. In general they are not interested in saving money, but are interested in saving critical materials, in cutting down personnel, and in doing nothing which would attract Congressional attention to our hi-jinks. I found that there were 1800 people working at the school to get things built fast.

With all good greetings—

Robert Oppenheimer

QUESTION	ANSWER
I. Housing	
1. Type of house	2 and 4-family units; bachelor quarters

QUESTION	ANSWER
Number of rooms	3, 4, 6 rooms (1, 2, 3 bedrooms)
Size of rooms	Kitchen, 10 × 14; livingroom 14 × 18
2. Type of heat	Hot air—one furnace for unit; wood fuel
Hot water heater	One heater for unit; wood fuel
Type of stove	Insulated wood & coal range, electric plate
Electricity?	Yes
Phone?	No
3. What furniture will be furnished?	Shelves, kitchen table, beds, chairs . . .
Ice box	Electric refrigerator
Ironing boards	Probably
Fireplace	Yes
4. Rent?	Free for Army;? for civilians*
What price for utilities?	Free for Army;? for civilians*
5. Will there be servants available for occasional help, especially for heavy work?	Yes
6. Small garden?	Yes; water limited in Spring and Fall
II. Geography and clothing	
1. Will it be permissible to describe location in letter?	Altitude 7300, in Southwest
2. Mean temperatures	Monthly means from 27.1 in January to 66.7 in July
Rain and snow fall	Heavy rain in summer; light snow in winter; temperate summers
3. Type of clothing needed	Warm clothing; informal
4. Shoes	Boots for occasional use; skis
III. Community Services	
1. Public laundry charge	Probably a small charge
2. Garbage collection	Yes

*For Army, most services free. For civilians, may be included in rent to make $150/mo. subsistence. Not yet decided.

QUESTION	ANSWER
3. Carpenters and other crafts	Yes
4. Store	Army Post Exchange
5. Hospital	Run by doctor
6. Mess hall or cafeteria	Mess hall and care
7. City tax	Probably not

C. Memorandum to Leslie R. Groves

Los Alamos
April 30, 1943

In accordance with our discussion of last week, I have given some thought to the question of a story about the Los Alamos Project which, if disseminated in the proper way, might serve somewhat to reduce the curiosity of the local population, and at least to delay the dissemination of the truth.

We propose that it be let known that the Los Alamos Project is working on a new type of rocket and that the detail be added that this is a largely electrical device. We feel that the story will have a certain credibility; that the loud noises which we will soon be making here will fit in with the subject; and that the fact, unfortunately not kept completely secret, that we are installing a good deal of electrical equipment, and the further fact that we have a large group of civilian specialists would fit in quite well. We further believe that the remoteness of the site for such a development and the secrecy which has surrounded the project would both be appropriate, and that the circumstance that a good deal of work is in fact being done on rockets, together with the appeal of the word, makes this story one which is both exciting and credible.

This question has been discussed with the governing board of the laboratory who approve it and who further recommend that the technical staff of the laboratory be specially warned neither to contradict nor to support a story of this kind if they should run into it.

J. R. Oppenheimer

D. Leslie R. Groves to Robert Oppenheimer

July 29, 1943

Dear Dr. Oppenheimer:

In view of the nature of the work on which you are engaged, the knowledge of it which is possessed by you and the dependence which rests upon you for its successful accomplishment, it seems necessary to ask you to take certain special precautions with respect to your personal safety.

It is requested that:

(a) You refrain from flying in airplanes of any description; the time saved is not worth the risk. (If emergency demands their use my prior consent should be requested.)

(b) You refrain from driving an automobile for any appreciable distance (above a few miles) and from being without suitable protection on any lonely road, such as the road from Los Alamos to Santa Fe. On such trips you should be accompanied by a competent, able bodied, armed guard. There is no objection to the guard serving as chauffeur.

(c) Your cars be driven with due regard to safety and that in driving about town a guard of some kind should be used, particularly during hours of darkness. The cost of such guard is a proper charge against the United States.

I realize that these precautions may be personally burdensome and that they may appear to you to be unduly restrictive but I am asking you to bear with them until our work is successfully completed.

Sincerely,
L. R. Groves
Brigadier General, C.E.

E. To Leslie R. Groves

Los Alamos
November 2, 1943

Dear General Groves:

After you gave me the list during your last visit of the men whom we may expect from the United Kingdom, it occurred to me that it might be wise before they arrive here to give them new names. This refers especially to Niels Bohr. I am thinking of the fact that mail will be addressed to them, that they may on occasion originate or receive long-distance calls, that they will be making some local purchases, and that for all these routine matters it would be preferable if such well known names were not put in circulation.

It has, in fact, troubled us some that we are forced to place calls for Dr. Conant, Fermi, Lawrence, etc. This does not happen very often, but in view of the fact that we try not to use these names over the telephone, the placing of the calls themselves seems to us rather unwise. I doubt whether at this late date it would be practicable to assign new names to those who have been associated with the project in the past. In the case of Bohr and Chadwick I think it would be advisable to do so before they get here.

Sincerely yours,

J. R. Oppenheimer

F. To Leslie R. Groves

(Los Alamos)

January 17, 1944

Dear General Groves:

Dr. Baker (Niels Bohr) has left today and I think it appropriate to report to you briefly on his visit.

1. On the technical side Dr. Baker concerned himself primarily with the correlation and interpretation of the many new data on nuclear fission and related topics which have been obtained by this project. He left with us a brief report on the theoretical understanding of these data. It has been the point of view of this laboratory that in matters of such great importance, and where theories were involved which were new and unproven, all important quantities would have to be determined by experimental measurement, and I believe this policy was and is sound. Nevertheless, the advantage of some theoretical insight into the phenomena is very great indeed in that it enables us to evaluate experiments critically, to determine the relative priority of experiments, and in general to reduce the amount of futile discussion and waste motion. For all these reasons the work that Baker did for us should prove of very great value in the months to come. Baker concerned himself very little with the engineering problems of our program although he is of course aware of their importance and their difficulty.

2. By arrangement with Chadwick and me, Baker is to remain on the British payroll and all his expenses are to be paid by the British. A change in this arrangement will only be made if you or Dr. Chadwick see strong reasons to alter it.

3. I should like in a formal way to express my hope that Baker's collaboration with this project will continue, since it has been of great help to us and is likely to be so throughout the year.

4. By word and deed Dr. Baker has done everything he could to sup-

port this project and to indicate that he is sympathetic not only with its purposes and general method of procedure, but with the policies and achievements of the project's overall direction. I should like to make it quite clear that the effect of his presence on the morale of those with whom he came in contact was always positive and always helpful, and that I see every reason to anticipate that this will be true in the future.

Very sincerely yours,
J. R. Oppenheimer

G. To Leslie R. Groves

(Los Alamos)
February 14, 1944

Dear General Groves:

Dr. Peierls spent two and a half days at Site Y. Almost all of this time was taken up by conferences with members of the Theoretical Division and with the other members of the British mission.

1. One of the main purposes of Peierls' visit was to reach a decision on what parts of the work now being carried out in Britain should be continued. There was therefore a rather careful comparison of the methods and results of the British workers with ours, in so far as the British have been working on the same problems. No important discrepancies in results were revealed by this comparison. It was agreed that British analytical work on critical masses, multiplication rates and efficiencies could best be discontinued.

It was further agreed that their work on the optical analogy, although probably not of a character to justify its initiation at this time, might profitably be continued and should preferably be transferred to this laboratory. This would mean bringing Marshall, who has been in charge of the work in Britain to this laboratory.[1] We should have him work under the general supervision of O. R. Frisch. Plans to effect this have been initiated.

It was further agreed that the detailed numerical calculations on the blast wave now in progress in Britain should be continued for some time. These are more laborious and less perspicuous than the methods used here, but should provide a reliable check.

2. The chief contribution made by Peierls to the work of this laboratory lay in the following. The methods used by the British for integrating the

1. Frisch was already at Los Alamos. D. G. Marshall soon joined the British mission there.

blast wave equations in all their complexity are applicable to the physically different but formally similar problem of the hydrodynamics of implosion. Since I am not satisfied with the adequacy of the methods now in use in this laboratory for studying this latter problem, we went into the technical aspects of the British methods in considerable detail with Peierls, and are planning to attack the implosion problem along these lines with the highest possible urgency.

Sincerely yours,
J. R. Oppenheimer

H. To Leslie R. Groves

(Los Alamos)
August 31, 1944

Dear General Groves:

I should like you to have a report on the status of our X-10 stocks. Our records show, according to Dr. Kennedy's recent study, that we have received a total of 51 grams of this material. The material has been used for approximately 2500 separate experiments. The overall loss per experiment has been about 1 per cent. The result is that we now have in stock or accounted for by loan to other laboratories, 36 of the original 51 grams.

I believe that on the whole the mastery of the techniques necessary for handling the product which we have obtained in the last months has justified the use to which the material has been put and the losses incurred therein. This is also the opinion of Dr. Kennedy. Some further loss is to be anticipated during the experimental phase of our studies of this material, and perhaps another 5 or 10 grams will disappear in this way, although I confidently hope that this is an overestimate. On the other hand, we are now in a position to carry through the operations necessary for final fabrication with a very high yield (99%) and to recover almost all that is not included in the yield. I therefore believe that the material which we have dissipated will be paid for many times over by the effectiveness with which we can deal with production lots when they become available.

Sincerely yours,
J. R. Oppenheimer

I. To Leslie R. Groves

(Los Alamos)
October 6, 1944

Dear General Groves:

I am glad to transmit the enclosed report of Captain Parsons, with the general intent and spirit of which I am in full sympathy. There are a few points on which my evaluation differs somewhat from that expressed in the report and it seems appropriate to mention them at this time.

1. I believe that Captain Parsons somewhat misjudges the temper of the responsible members of the laboratory. It is true that there are a few people here whose interests are exclusively "scientific" in the sense that they will abandon any problem that appears to be soluble. I believe that these men are now in appropriate positions in the organization. For the most part the men actually responsible for the prosecution of the work have proven records of carrying developments through the scientific and into the engineering stage. For the most part these men regard their work here not as a scientific adventure, but as a responsible mission which will have failed if it is let drop at the laboratory phase. I therefore do not expect to have to take heroic measures to insure something which I know to be the common desire of the overwhelming majority of our personnel.

2. I agree completely with all the comments of Captain Parsons' memorandum on the fallacy of regarding a controlled test as the culmination of the work of this laboratory. The laboratory is operating under a directive to produce weapons; this directive has been and will be rigorously adhered to. The only reason why we contemplate making a test, and why I have in the past advocated this, is because with the present time scales and the present radical assembly design this appears to be a necessary step in the production of a weapon. I do not wish to prejudge the issue: it is possible that information available to us within the next months may make such a test unnecessary. I believe, however, that the probability of this is extremely small.

3. The developmental program of the laboratory, whether or not it has been prosecuted with intelligence and responsibility, is still far behind the minimal requirements set by our directive. This fact, which rests on no perfectionist ideals for long-range development, means that there must inevitably be some duplication of effort and personnel if the various phases of our program,—scientific, engineering and military,—are to be carried out without too great mutual interference. It is for this reason that I should like to stress Captain Parsons' remark that a very great strengthening in engi-

neering is required. The organizational experience which the last year has given us is no substitute for competent engineers.

Sincerely yours,
J. R. Oppenheimer

J. To the Secretary of War

August 17, 1945

Dear Mr. Secretary:

The Interim Committee has asked us to report in some detail on the scope and program of future work in the field of atomic energy. One important phase of this work is the development of weapons; and since this is the problem which has dominated our war time activities, it is natural that in this field our ideas should be most definite and clear, and that we should be most confident of answering adequately the questions put to us by the committee. In examining these questions we have, however, come on certain quite general conclusions, whose implications for national policy would seem to be both more immediate and more profound than those of the detailed technical recommendations to be submitted. We, therefore, think it appropriate to present them to you at this time.

1. We are convinced that weapons quantitatively and qualitatively far more effective than now available will result from further work on these problems. This conviction is motivated not alone by analogy with past developments, but by specific projects to improve and multiply the existing weapons, and by the quite favorable technical prospects of the realization of the super bomb.

2. We have been unable to devise or propose effective military countermeasures for atomic weapons. Although we realize that future work may reveal possibilities at present obscure to us, it is our firm opinion that no military countermeasures will be found which will be adequately effective in preventing the delivery of atomic weapons.

The detailed technical report in preparation will document these conclusions, but hardly alter them.

3. We are not only unable to outline a program that would assure to this nation for the next decades hegemony in the field of atomic weapons; we are equally unable to insure that such hegemony, if achieved, could protect us from the most terrible destruction.

4. The development, in the years to come, of more effective atomic weapons, would appear to be a most natural element in any national policy of maintaining our military forces at great strength; nevertheless we have grave doubts that this further development can contribute essentially or permanently to the prevention of war. We believe that the safety of this nation—as opposed to its ability to inflict damage on an enemy power—cannot lie wholly or even primarily in its scientific or technical prowess. It can be based only on making future wars impossible. It is our unanimous and urgent recommendation to you that, despite the present incomplete exploitation of technical possibilities in this field, all steps be taken, all necessary international arrangements be made, to this one end.

5. We should be most happy to have you bring these views to the attention of other members of the Government, or of the American people, should you wish to do so.

Very sincerely,
J. R. Oppenheimer
For the Panel

---

7. The Quebec Agreement, August 19, 1943

The Citadel, Quebec.

Articles of Agreement Governing Collaboration Between The
Authorities of the U.S.A. and the U.K.
in the Matter of Tube Alloys[1]

Whereas it is vital to our common safety in the present War to bring the Tube Alloys project to fruition at the earliest moment; and whereas this

This text appeared in Foreign Relations of the United States, *The First Quebec Conference* (Washington, D.C., 1970), pp. 1117–19; Margaret Gowing, *Britain and Atomic Energy 1939–1945* (New York: MacMillan, 1964), pp. 439–40.

1. These Articles of Agreement are typed on four pages of stationery on each of which appears the letterhead "The Citadel Quebec." For a photocopy of the British signed original of these Articles, see "Articles of Agreement governing collaboration between the authorities of the U.S.A. and U.K. in the matter of Tube Alloys" (Cmd. 9123; London: H.M. Stationery Office, 1954). The text of the articles printed by the Department of State in 1954 as Treaties and Other International Acts Series No. 2993 was prepared from a photocopy of the British signed original. The United States signed original used as the source text for the document printed

may be more speedily achieved if all available British and American brains and resources are pooled; and whereas owing to war conditions it would be an improvident use of war resources to duplicate plants on a large scale on both sides of the Atlantic and therefore a far greater expense has fallen upon the United States;

It is agreed between us

First, that we will never use this agency against each other.

Secondly, that we will not use it against third parties without each other's consent.

Thirdly, that we will not either of us communicate any information about Tube Alloys to third parties except by mutual consent.

Fourthly, that in view of the heavy burden of production falling upon the United States as the result of a wise division of war effort, the British Government recognize that any post-war advantages of an industrial or commercial character shall be dealt with as between the United States and Great Britain on terms to be specified by the President of the United States to the Prime Minister of Great Britain. The Prime Minister expressly disclaims any interest in these industrial and commercial aspects beyond what may be considered by the President of the United States to be fair and just and in harmony with the economic welfare of the world.

And Fifthly, that the following arrangements shall be made to ensure full and effective collaboration between the two countries in bringing the project to fruition:

(a) There shall be set up in Washington a Combined Policy Committee composed of:

The Secretary of War.	(United States)
Dr. Vannevar Bush.	(United States)
Dr. James B. Conant.	(United States)

---

here is identical with the British signed original except that (a) the three United States members of the Combined Policy Committee appear in typed form in the United States original whereas they are in Roosevelt's handwriting in the British original; (b) the form of the date at the end of the document (in Roosevelt's handwriting on both copies) reads "August 19 1943" in the British original and "Aug. 19th 1943" in the United States original (in which also the first digit of 19th appears to have been written over a figure 2, presumably because Roosevelt began to write 20th and then corrected it to 19th); and (c) there is a period in the United States original after the surname of C. D. Howe.

The source text of the Articles of Agreement is attached to a memorandum from Churchill's Principal Private Secretary to Roosevelt's Naval Aide, dated at Quebec, August 19, 1943, which reads as follows: "Admiral Wilson Brown. I attach, for retention, one of the two copies of the Articles of Agreement relating to Tube Alloys, signed by the President and Mr. Churchill today. J. M. Martin."

Field-Marshal Sir John Dill, G.C.B.,
    C.M.G., D.S.O.   (United Kingdom)
Colonel the Right Hon. J. J. Llewellin,
    C.B.E., M.C., M.P.   (United Kingdom)
The Honourable C. D. Howe.   (Canada)

The functions of this Committee, subject to the control of the respective Governments, will be:

(1) To agree from time to time upon the programme of work to be carried out in the two countries.

(2) To keep all sections of the project under constant review.

(3) To allocate materials, apparatus and plant, in limited supply, in accordance with the requirements of the programme agreed by the Committee.

(4) To settle any questions which may arise on the interpretation or application of this Agreement.

(b) There shall be complete interchange of information and ideas on all sections of the project between members of the Policy Committee and their immediate technical advisers.

(c) In the field of scientific research and development there shall be full and effective interchange of information and ideas between those in the two countries engaged in the same sections of the field.

(d) In the field of design, construction and operation of large-scale plants, interchange of information and ideas shall be regulated by such ad hoc arrangements as may, in each section of the field, appear to be necessary or desirable if the project is to be brought to fruition at the earliest moment. Such ad hoc arrangements shall be subject to the approval of the Policy Committee.

Approved[2]

Aug. 19th 1943[2]   Franklin D. Roosevelt
                                 Winston S. Churchill

2. In Roosevelt's handwriting in the source text.

8. Anglo-American Declaration of Trust, June 13, 1944

This Agreement and Declaration of Trust is made the thirteenth day of June One thousand nine hundred and fourty four by Franklin Delano Roosevelt on behalf of the Government of the United States of America, and by Winston Leonard Spencer Churchill on behalf of the Government of the United Kingdom of Great Britain and Northern Ireland. The said Governments are hereinafter referred to as "the Two Governments";

Whereas an agreement (hereinafter called the Quebec Agreement) was entered into on the nineteenth day of August One thousand nine hundred and forty three by and between the President of the United States and the Prime Minister of the United Kingdom and

Whereas it is an object vital to the common interests of those concerned in the successful prosecution of the present war to insure the acquisition at the earliest practicable moment of an adequate supply of uranium and thorium ores; and

Whereas it is the intention of the Two Governments to control to the fullest extent practicable the supplies of uranium and thorium ores within the boundaries of such areas as come under their respective jurisdictions; and

Whereas the Government of the United Kingdom of Great Britain and Northern Ireland intends to approach the Governments of the Dominions and the Governments of India and of Burma for the purpose of securing that such Governments shall bring under control deposits of the uranium and thorium ores within their respective territories; and

Whereas it has been decided to establish a joint organisation for the purpose of gaining control of the uranium and thorium supplies in certain areas outside the control of the Two Governments and of the Governments of the Dominions and of India and of Burma;

Now it is Hereby Agreed and Declared as Follows:

1. (1) There shall be established in the City of Washington, District of Columbia, a Trust to be known as "The Combined Development Trust".
   (2) The Trust shall be composed of and administered by six persons who shall be appointed, and the subject to removal, by the Combined Policy Committee established by the Quebec Agreement.
2. The Trust shall use its best endeavours to gain control of and develop the production of the uranium and thorium supplies situate in certain

This text appeared in Margaret Gowing, *Britain and Atomic Energy 1939–1945* (New York: MacMillan, 1964), pp. 441–42.

areas other than the areas under the jurisdiction of the Two Governments and of the Governments of the Dominions and of India and of Burma and for that purpose shall take such steps as it may in the common interest think fit to:

a Explore and survey sources of uranium and thorium supplies.

b Develop the production of uranium and thorium by the acquisition of mines and ore deposits, mining concessions or otherwise.

c Provide with equipment any mines or mining works for the production of uranium and thorium.

d Survey and improve the methods of production of uranium and thorium.

e Acquire and undertake the treatment and disposal of uranium and thorium and uranium and thorium materials.

f Provide storage and other facilities.

g Undertake any functions or operations which conduce to the effective carrying out of the purpose of the Trust in the common interest.

3. (1) The Trust shall carry out its functions under the direction and guidance of the Combined Policy Committee, and as its agent, and all uranium and thorium ores and supplies and other property acquired by the Trust shall be held by it in trust for the Two Governments jointly, and disposed of or otherwise dealt with in accordance with the direction of the Combined Policy Committee.

(2) The Trust shall submit such reports of its activities as may be required from time to time by the Combined Policy Committee.

4. For the purpose of carrying out its functions, the Trust shall utilize whenever and wherever practicable the established agencies of any of the Two Governments, and may employ and pay such other agents and employees as it considers expedient, and may delegate to any agents or employees all or any of its functions.

5. The Trust may acquire and hold any property in the name of nominees.

6. All funds properly required by the Trust for the performance of its functions shall be provided as to one-half by the Government of the United States of America and the other half by the Government of the United Kingdom of Great Britain and Northern Ireland.

7. In the event of the Combined Policy Committee ceasing to exist, the functions of the Committee under the Trust shall be performed by such other body or person as may be designated by the President for the time being of the United States of America and the Prime Minister for the time being of the United Kingdom of Great Britain and Northern Ireland.

8. The signatories of this Agreement and Declaration of Trust will, as soon

as practicable after the conclusion of hostilities, recommend to their respective Governments the extension and revision of this war-time emergency agreement to cover post war conditions and its formalization by
treaty or other proper method. This Agreement and Declaration of
Trust shall continue in full force and effect until such extension or
revision.

(Signed) Franklin D. Roosevelt
On Behalf of the Government of the United States of
America

(Signed) Winston S. Churchill
On Behalf of the Government of the United Kingdom of
Great Britain and Northern Ireland

---

9. Roosevelt-Churchill Hyde Park Aide-Memoire, September 19, 1944

1. The suggestion that the world should be informed regarding tube
alloys, with a view to an international agreement regarding its control and
use, is not accepted. The matter should continue to be regarded as of the
utmost secrecy; but when a "bomb" is finally available, it might perhaps,
after mature consideration, be used against the Japanese, who should be
warned that this bombardment will be repeated until they surrender.

2. Full collaboration between the United States and the British Government in developing tube alloys for military and commercial purposes
should continue after the defeat of Japan unless and until terminated by
joint agreement.

3. Enquiries should be made regarding the activities of Professor Bohr
and steps taken to ensure that he is responsible for no leakage of information particularly to the Russians.

This text appeared in Foreign Relations of the United States, *1944*, vol. 2 (Washington, D.C.,
1967), pp. 1026–28.

10. O. R. Frisch, Eyewitness Account of "Trinity" Test, July 1945

I watched the explosion from a point said to be about 20 (or 25) miles away and about north of it, together with the members of the co-ordinating council. Fearing to be dazzled and to be burned by ultraviolet rays, I stood with my back to the gadget, and behind the radio truck. I looked at the hills, which were visible in the first faint light of dawn (0530 M.W. Time). Suddenly and without any sound, the hills were bathed in brilliant light, as if somebody had turned the sun on with a switch. It is hard to say whether the light was less or more brilliant than full sunlight, since my eyes were pretty well dark adapted. The hills appeared kind of flat and colourless like a scenery seen by the light of a photographic flash, indicating presumably that the retina was stimulated beyond the point where intensity discrimination is adequate. The light appeared to remain constant for about one or two seconds (probably for the same reason) and then began to diminish rapidly.

After that I turned round and tried to look at the light source but found it still too bright to keep my eyes on it. A few short glances gave me the impression of a small very brilliant core much smaller in appearance than the sun, surrounded by decreasing and reddening brightness with no definite boundary, but not greater than the sun. After some seconds I could keep my eye on the thing and it now looked like a pretty perfect red ball, about as big as the sun, and connected to the ground by a short grey stem. The ball rose slowly, lengthening its stem and getting gradually darker and slightly larger. A structure of darker and lighter irregularities became visible, making the ball look somewhat like a raspberry. Then its motion slowed down and it flattened out, but still remained connected to the ground by its stem, looking more than ever like the trunk of an elephant. Then a hump grew out of its top surface and a second mushroom grew out of the top of the first one, slowly penetrating the highest cloud layers. As the red glow died out it became apparent that the whole structure, in particular the top mushroom, was surrounded by a purplish blue glow. A minute or so later the whole top mushroom appeared to glow feebly in this colour, but this was no longer easy to see, in the increasing light of dawn.

A very striking phenomenon was the sudden appearance of a white patch on the underside of the cloud layer just above the explosion; the patch spread very rapidly, like a pool of spilt milk, and a second or two

This text appeared in Foreign Relations of the United States, *Potsdam*, vol. 2 (Washington, D.C., 1960) p. 1371.

later, a similar patch appeared and spread on another cloud layer higher up. They marked no doubt the impact of the blast wave on the cloud layers. They appeared, I believe, before the red ball had started to flatten out.

When I thought it was soon time for the blast to arrive, I sat on the ground, still facing the explosion, and put my fingers in my ears. Despite that, the report was quite respectable and was followed by a long rumbling, not quite like thunder but more regular, like huge noisy waggons running around in the hills.

11. General Groves's Report on "Trinity," July 18, 1945

TOP SECRET   WASHINGTON, 18 July 1945.

Memorandum for the Secretary of War[1]

Subject: The Test.

1. This is not a concise, formal military report but an attempt to recite

This text appeared in Foreign Relations of the United States, *Potsdam*, vol. 2 (Washington, D.C., 1960), pp. 1361–68.

1. [Henry] Stimson's diary entry for July 21 contains the following information relating to this document:

At eleven thirty-five General Groves' special report was received by special courier. It was an immensely powerful document, clearly and well written and with supporting documents of the highest importance. It gave a pretty full and eloquent report of the tremendous success of the test and revealed far greater destructive power than we expected in S-1. . . .

At three o'clock I found that Marshall had returned from the Joint Chiefs of Staff, and to save time I hurried to his house and had him read Groves' report and conferred with him about it.

I then went to the "Little White House" and saw President Truman. I asked him to call in Secretary Byrnes and then I read the report in its entirety and we then discussed it. They were immensely pleased. The President was tremendously pepped up by it and spoke to me of it again and again when I saw him. He said it gave him an entirely new feeling of confidence and he thanked me for having come to the Conference and being present to help him in this way.

Stimson showed Groves's report to [Gen. H. H.] Arnold on July 22.

Truman later stated that, following receipt of news that the Alamogordo test had been successful, he had called together [Sec. James F.] Byrnes, Stimson, [Adm. William D.] Leahy, [Gen. George] Marshall, Arnold, [Gen. Dwight] Eisenhower, and [Adm. Ernest J.] King and had asked them for their opinions as to whether the bomb should be used, and the consensus had been that it should . . . Truman apparently also received at this meeting an oral estimate of the casualties to be expected in the assault of Japan if the new weapon were not used.

what I would have told you if you had been here on my return from New Mexico.

2. At 0530,[2] 16 July 1945, in a remote section of the Alamogordo Air Base, New Mexico, the first full scale test was made of the implosion of the atomic fission bomb. For the first time in history there was a nuclear explosion. And what an explosion! . . . The bomb was not dropped from an airplane but was exploded on a platform on top of a 100-foot high steel tower.

3. The test was successful beyond the most optimistic expectations of anyone. Based on the data which it has been possible to work up to date, I estimate the energy generated to be in excess of the equivalent of 15,000 to 20,000 tons of TNT; and this is a conservative estimate. Data based on measurements which we have not yet been able to reconcile would make the energy release several times the conservative figure. There were tremendous blast effects. For a brief period there was a lighting effect within a radius of 20 miles equal to several suns in midday; a huge ball of fire was formed which lasted for several seconds. This ball mushroomed and rose to a height of over ten thousand feet before it dimmed. The light from the explosion was seen clearly at Albuquerque, Santa Fe, Silver City, El Paso and other points generally to about 180 miles away. The sound was heard to the same distance in a few instances but generally to about 100 miles. Only a few windows were broken although one was some 125 miles away. A massive cloud was formed which surged and billowed upward with tremendous power, reaching the substratosphere at an elevation of 41,000 feet, 36,000 feet above the ground in about five minutes, breaking without interruption through a temperature inversion at 17,000 feet which most of the scientists thought would stop it. Two supplementary explosions occurred in the cloud shortly after the main explosion. The cloud contained several thousand tons of dust picked up from the ground and a considerable amount of iron in the gaseous form. Our present thought is that this iron ignited when it mixed with the oxygen in the air to cause these supplementary explosions. Huge concentrations of highly radioactive materials resulted from the fission and were contained in this cloud.

4. A crater from which all vegetation had vanished, with a diameter of 1200 feet and a slight slope toward the center, was formed. In the center was a shallow bowl 130 feet in diameter and 6 feet in depth. The material within the crater was deeply pulverized dirt. The material within the outer circle is greenish and can be distinctly seen from as much as 5 miles away.

---

2. I.e., 5:30 A.M. All times in this memorandum are expressed in military style, i.e., from 0001 hours (12:01 A.M.) to 2400 hours (midnight).

The steel from the tower was evaporated. 1500 feet away there was a four-inch iron pipe 16 feet high set in concrete and strongly guyed. It disappeared completely.

5. One-half mile from the explosion there was a massive steel test cylinder weighing 220 tons. The base of the cylinder was solidly encased in concrete. Surrounding the cylinder was a strong steel tower 70 feet high, firmly anchored to concrete foundations. This tower is comparable to a steel building bay that would be found in typical 15 or 20 story skyscraper or in warehouse construction. Forty tons of steel were used to fabricate the tower which was 70 feet high, the height of a six story building. The cross bracing was much stronger than normally used in ordinary steel construction. The absence of the solid walls of a building gave the blast a much less effective surface to push against. The blast tore the tower from its foundations, twisted it, ripped it apart and left it flat on the ground. The effects on the tower indicate that, at that distance, unshielded permanent steel and masonry buildings would have been destroyed. I no longer consider the Pentagon a safe shelter from such a bomb. Enclosed are a sketch showing the tower before the explosion and a telephotograph showing what it looked like afterwards.[3] None of us had expected it to be damaged.

6. The cloud traveled to a great height first in the form of a ball, then mushroomed, then changed into a long trailing chimney-shaped column and finally was sent in several directions by the variable winds at the different elevations. It deposited its dust and radioactive materials over a wide area. It was followed and monitored by medical doctors and scientists with instruments to check its radioactive effects. While here and there the activity on the ground was fairly high, at no place did it reach a concentration which required evacuation of the population. Radioactive material in small quantities was located as much as 120 miles away. The measurements are being continued in order to have adequate data with which to protect the Government's interests in case of future claims. For a few hours I was none too comfortable about the situation.

7. For distances as much as 200 miles away, observers were stationed to check on blast effects, property damage, radioactivity and reactions of the population. While complete reports have not yet been received, I now know that no persons were injured nor was there any real property damage outside our Government area. As soon as all the voluminous data can be checked and correlated, full technical studies will be possible.

8. Our long range weather predictions had indicated that we could expect weather favorable for our tests beginning on the morning of the 17th

3. Neither reproduced.

and continuing for four days. This was almost a certainty if we were to believe our long range forecasters. The prediction for the morning of the 16th was not so certain but there was about an 80% chance of the conditions being suitable. During the night there were thunder storms with lightning flashes all over the area. The test had been originally set for 0400 hours and all the night through, because of the bad weather, there were urgings from many of the scientists to postpone the test. Such a delay might well have had crippling results due to mechanical difficulties in our complicated test set-up. Fortunately, we disregarded the urgings. We held firm and waited the night through hoping for suitable weather. We had to delay an hour and a half, to 0530, before we could fire. This was 30 minutes before sunrise.

9. Because of bad weather, our two B-29 observation airplanes were unable to take off as scheduled from Kirtland Field at Albuquerque and when they finally did get off, they found it impossible to get over the target because of the heavy clouds and the thunder storms. Certain desired observations could not be made and while the people in the airplanes saw the explosion from a distance, they were not as close as they will be in action. We still have no reason to anticipate the loss of our plane in an actual operation although we cannot guarantee safety.

10. Just before 1100 the news stories from all over the state started to flow into the Albuquerque Associated Press. I then directed the issuance by the Commanding Officer, Alamogordo Air Base of a news release as shown on the enclosure.[4] With the assistance of the Office of Censorship we were able to limit the news stories to the approved release supplemented in the local papers by brief stories from the many eyewitnesses not connected with our project. One of these was a blind woman who saw the light.

11. Brigadier General Thomas F. Farrell was at the control shelter lo-

4. Identified in the source copy as a clipping from *The Albuquerque Tribune* for July 16, 1945:

Alamogordo, N. M., July 16—William O. Eareckson, commanding officer of the Alamogordo Army Air Base, made the following statement today:

"Several inquiries have been received concerning a heavy explosion which occurred on the Alamogordo Air Base reservation this morning.

"A remotely located ammunition magazine containing a considerable amount of high explosive and pyrotechnics exploded.

"There was no loss of life or injury to anyone, and the property damage outside of the explosives magazine itself was negligible.

"Weather conditions affecting the content of gas shells exploded by the blast may make it desirable for the Army to evacuate temporarily a few civilians from their homes."

cated 10,000 yards south of the point of explosion. His impressions are
given below:

"The scene inside the shelter was dramatic beyond words. In and
around the shelter were some twenty-odd people concerned with last
minute arrangements prior to firing the shot. Included were: Dr. Oppenheimer, the Director who had borne the great scientific burden of developing the weapon from the raw materials made in Tennessee and Washington
and a dozen of his key assistants—Dr. Kistiakowsky, who developed the
highly special explosives; Dr. Bainbridge, who supervised all the detailed
arrangements for the test; Dr. Hubbard, the weather expert, and several
others. Besides these, there were a handful of soldiers, two or three Army
officers and one Naval officer. The shelter was cluttered with a great variety
of instruments and radios.

"For some hectic two hours preceding the blast, General Groves stayed
with the Director, walking with him and steadying his tense excitement.
Every time the Director would be about to explode because of some un-
toward happening General Groves would take him off and walk with him
in the rain, counselling with him and reassuring him that everything would
be all right. At twenty minutes before zero hours, General Groves left for
his station at the base camp, first because it provided a better observation
point and second, because of our rule that he and I must not be together in
situations where there is an element of danger, which existed at both
points.

"Just after General Groves left, announcements began to be broadcast
of the interval remaining before the blast. They were sent by radio to the
other groups participating in and observing the test. As the time interval
grew smaller and changed from minutes to seconds, the tension increased
by leaps and bounds. Everyone in that room knew the awful potentialities
of the thing that they thought was about to happen. The scientists felt that
their figuring must be right and that the bomb had to go off but there was
in everyone's mind a strong measure of doubt. The feeling of many could
be expressed by 'Lord, I believe; help Thou mine unbelief.' We were
reaching into the unknown and we did not know what might come of it. It
can be safely said that most of those present—Christian, Jew and Atheist—
were praying and praying harder than they had ever prayed before. If the
shot were successful, it was a justification of the several years of intensive
effort of tens of thousands of people—statesmen, scientists, engineers,
manufacturers, soldiers, and many others in every walk of life.

"In that brief instant in the remote New Mexico desert the tremendous
effort of the brains and brawn of all these people came suddenly and star-

tlingly to the fullest fruition. Dr. Oppenheimer, on whom had rested a very heavy burden, grew tenser as the last seconds ticked off. He scarcely breathed. He held on to a post to steady himself. For the last few seconds, he stared directly ahead and then when the announcer shouted 'Now!' and there came this tremendous burst of light followed shortly thereafter by the deep growling roar of the explosion, his face relaxed into an expression of tremendous relief. Several of the observers standing back of the shelter to watch the lighting effects were knocked flat by the blast.

"The tension in the room let up and all started congratulating each other. Everyone sensed 'This is it!' No matter what might happen now all knew that the impossible scientific job had been done. Atomic fission would no longer be hidden in the cloisters of the theoretical physicists' dreams. It was almost full grown at birth. It was a great new force to be used for good or for evil. There was a feeling in that shelter that those concerned with its nativity should dedicate their lives to the mission that it would always be used for good and never for evil.

"Dr. Kistiakowsky,[5] the impulsive Russian, threw his arms around Dr. Oppenheimer and embraced him with shouts of glee. Others were equally enthusiastic. All the pent-up emotions were released in those few minutes and all seemed to sense immediately that the explosion had far exceeded the most optimistic expectations and wildest hopes of the scientists. All seemed to feel that they had been present at the birth of a new age—The Age of Atomic Energy—and felt their profound responsibility to help in guiding into right channels the tremendous forces which had been unlocked for the first time in history.

"As to the present war, there was a feeling that no matter what else might happen, we now had the means to insure its speedy conclusion and save thousands of American lives. As to the future, there had been brought into being something big and something new that would prove to be immeasurably more important than the discovery of electricity or any of the other great discoveries which have so affected our existence.

"The effects could well be called unprecedented, magnificent, beautiful, stupendous and terrifying. No man-made phenomenon of such tremendous power had ever occurred before. The lighting effects beggared description. The whole country was lighted by a searing light with the intensity many times that of the midday sun. It was golden, purple, violet, gray and blue. It lighted every peak, crevasse and ridge of the nearby mountain range with a clarity and beauty that cannot be described but must be seen to be imagined. It was that beauty the great poets dream

5. At this point is the following manuscript interpolation by Groves: "An American and Harvard professor for many years."

about but describe most poorly and inadequately. Thirty seconds after the explosion came first, the air blast pressing hard against the people and things, to be followed almost immediately by the strong, sustained, awesome roar which warned of doomsday and made us feel that we puny things were blasphemous to dare tamper with the forces heretofore reserved to The Almighty. Words are inadequate tools for the job of acquainting those not present with the physical, mental and psychological effects. It had to be witnessed to be realized."

12. My impressions of the night's high points follow:

After about an hour's sleep I got up at 0100 and from that time on until about five I was with Dr. Oppenheimer constantly. Naturally he was nervous, although his mind was working at its usual extraordinary efficiency. I devoted my entire attention to shielding him from the excited and generally faulty advice of his assistants who were more than disturbed by their excitement and the uncertain weather conditions. By 0330 we decided that we could probably fire at 0530. By 0400 the rain had stopped but the sky was heavily overcast. Our decision became firmer as time went on. During most of these hours the two of us journeyed from the control house out into the darkness to look at the stars and to assure each other that the one or two visible stars were becoming brighter. At 0510 I left Dr. Oppenheimer and returned to the main observation point which was 17,000 yards from the point of explosion. In accordance with our orders I found all personnel not otherwise occupied massed on a bit of high ground.

At about two minutes of the scheduled firing time all persons lay face down with their feet pointing towards the explosion. As the remaining time was called from the loud speaker from the 10,000 yard control station there was complete silence. Dr. Conant said he had never imagined seconds could be so long. Most of the individuals in accordance with orders shielded their eyes in one way or another. There was than this burst of light of a brilliance beyond any comparison. We all rolled over and looked through dark glasses at the ball of fire. About forty seconds later came the shock wave followed by the sound, neither of which seemed startling after our complete astonishment to the extraordinary lighting intensity. Dr. Conant reached over and we shook hands in mutual congratulations. Dr. Bush, who was on the other side of me, did likewise. The feeling of the entire assembly was similar to that described by General Farrell, with even the uninitiated feeling profound awe. Drs. Conant and Bush and myself were struck by an even stronger feeling that the faith of those who had been responsible for the initiation and the carrying on of this Herculean project had been justified. I personally thought of Blondin crossing Niagara Falls on his tight rope, only to me this tight rope had lasted for almost three years and of my repeated confident-appearing assurances that such a thing was possible and that we would do it.

13. A large group of observers were stationed at a point about 27 miles north of the point of explosion. Attached is a memorandum written shortly after the explosion by Dr. E. O. Lawrence which may be of interest.

14. While General Farrell was waiting about midnight for a commercial airplane to Washington at Albuquerque—120 miles away from the site—he overheard several airport employees discussing their reaction to the blast. One said that he was out on the parking apron; it was quite dark; the light lasted several seconds. Another remarked that if a few exploding bombs could have such an effect, it must be terrible to have them drop on a city.

15. My liaison officer at the Alamogordo Air Base, 60 miles away, made the following report:

"There was a blinding flash of light that lighted the entire northwestern sky. In the center of the flash, there appeared to be a huge billow of smoke. The original flash lasted approximately 10 to 15 seconds. As the first flash died down, there arose in the approximate center of where the original flash had occurred an enormous ball of what appeared to be fire and closely resembled a rising sun that was three-fourths above a mountain. The ball of fire lasted approximately 15 seconds, then died down and the sky resumed an almost normal appearance.

"Almost immediately, a third, but much smaller, flash and billow of smoke of a whitish-orange color appeared in the sky, again lighting the sky for approximately 4 seconds. At the time of the original flash, the field was lighted well enough so that a newspaper could easily have been read. The second and third flashes were of much lesser intensity.

"We were in a glass-enclosed control tower some 70 feet above the ground and felt no concussion or air compression. There was no noticeable earth tremor although reports overheard at the Field during the following 24 hours indicated that some believed that they had both heard the explosion and felt some earth tremor."

16. I have not written a separate report for General Marshall as I feel you will want to show this to him. I have informed the necessary people here of our results. Lord Halifax after discussion with Mr. Harrison and myself stated that he was not sending a full report to his government at this time. I informed him that I was sending this to you and that you might wish to show it to the proper British representatives.

17. We are all fully conscious that our real goal is still before us. The battle test is what counts in the war with Japan.

18. May I express my deep personal appreciation for your congratula-

tory cable to us and for the support and confidence which I have received from you ever since I have had this work under my charge.

19. I know that Colonel Kyle will guard these papers with his customary extraordinary care.

L. R. Groves

---

12. Truman-Stalin Conversation, Potsdam, July 24, 1945

Present[1]

United States[2]	*Soviet Union*
President Truman	Generalissimo Stalin
	Mr. Pavlov

[Harry] Truman gives the following account of the conversation in *Year of Decisions*, p. 416: "On July 24 I casually mentioned to Stalin that we had a new weapon of unusual destructive force. The Russian Premier showed no special interest. All he said was that he was glad to hear it and hoped we would make 'good use of it against the Japanese.' "

[James F.] Byrnes gives the following information on the conversation in *Speaking Frankly*, p. 263: "At the close of the meeting of the Big Three on the afternoon of July 24, the President walked around the large circular table to talk to Stalin. After a brief conversation the President rejoined me and we rode back to the 'Little White House' together. He said he had told Stalin that, after long experimentation, we had developed a new bomb far more destructive than any other known bomb, and that we planned to use it very soon unless Japan surrendered. Stalin's only reply was to say that he was glad to hear of the bomb and he hoped we would use it."

This text appeared in Foreign Relations of the United States, *Potsdam*, vol. 2 (Washington, D.C., 1960), pp. 1376–79.

1. This conversation took place at the Cecilienhof Palace immediately after the adjournment of the Eighth Plenary Meeting. [EDITOR'S NOTE: No official record of it has ever been found.]

2. Byrnes (*All in One Lifetime*, p. 300) mentions [Charles] Bohlen's presence as Truman's interpreter at this conversation, and, in a conference Department of State historians had with Truman and members of his staff on January 24, 1956, Truman supplied the information that both Bohlen and a Soviet interpreter were present at this conversation. In a conversation with a Department of State historian on January 26, 1960, however, Bohlen stated categorically that Truman walked over to Stalin alone in order to give the conversation a more casual flavor, and that Pavlov did the interpreting.

[William D.] Leahy's account is as follows (*I Was There*, p. 429): "At the plenary session on July 24, Truman walked around to Stalin and told him quietly that we had developed a powerful weapon, more potent than anything yet seen in war. The President said later that Stalin's reply indicated no especial interest and that the Generalissimo did not seem to have any conception of what Truman was talking about. It was simply another weapon and he hoped we would use it effectively."

[Winston] Churchill, who was also an eye-witness to the conversation, gives the following information (*Triumph and Tragedy*, pp. 669–70): "Next day, July 24, after our plenary meeting had ended and we all got up from the round table and stood about in twos and threes before dispersing, I saw the President go up to Stalin, and the two conversed alone with only their interpreters. I was perhaps five yards away, and I watched with the closest attention the momentous talk. I knew what the President was going to do. What was vital to measure was its effect on Stalin. I can see it all as if it were yesterday. He seemed to be delighted. . . . As we were waiting for our cars I found myself near Truman. 'How did it go?' I asked. 'He never asked a question,' he replied."

13. Military Policy Committee Minutes, May 5, 1943

The point of use of the first bomb was discussed and the general view appeared to be that its best point of use would be on a Japanese fleet concentration in the Harbor of Truk. General Styer suggested Tokio [*sic*], but it was pointed out that the bomb should be used where, if it failed to go off, it would land in water of sufficient depth to prevent easy salvage. The Japanese were selected as they would not be so apt to secure knowledge from it as would the Germans.

This document is found in Manhattan Engineer District-Top Secret, folder 23A, Record Group 77, National Archives, Washington, D.C.

14. Stimson and the Interim Committee, May 1945

War Department
Washington

1 May 1945

Memorandum for the Secretary of War
Subject: Interim Committee on S-1.

Last week you presented to President Truman a fairly complete memorandum on the S-1 project, outlining its genesis, its present state of development and in general its availability for military usage. Your presentation was accompanied by a brief memo which you prepared relative to the broader political and international implications of the problem and the need for post war controls, both national and international. You had in mind the advisability of setting up a committee of particular qualifications for recommending action to the executive and legislative branches of the government when secrecy is no longer fully required. The committee would also be expected to recommend actions to be taken by the War Department in anticipation of the post war problems.

In view of the possibly short time available before actual military use and the relaxation of secrecy, it seems to me,—and as you know both Dr. Conant and Dr. Bush agree—that it is becoming more and more important to organize such a committee as promptly as possible. This committee should, I think, be a relatively small committee which should be prepared to serve temporarily or until Congress might appoint a permanent Post War commission to supervise, regulate and control the use of the product.

Certain things, however, must be done now before use if we are to avoid the risk of grave repercussions on the public in general and on Congress in particular. For instance, the committee will need to prepare appropriate announcements to be available for issue (a) by the President and (b) by the War Department as soon as the first bomb is used. These announcements or later publicity would presumably give some of the history of the project, its importance from a military standpoint, its scientific background, and some of its dangers. Most importantly as soon as possible after use some assurance must be given of the steps to be taken to provide the essential controls over post-war use and development, both at home and abroad. With that in mind it will be necessary as soon as possible after use

This document is found in Manhattan Engineer District-Top Secret, Harrison-Bundy files, folder 69, National Archives, Washington, D.C.

to make recommendations for the necessary Congressional legislation covering patents, use, controls, etc.

All of these and many other factors will have to be studied by the committee with the understanding that all recommendations must be for your own approval and for submission to the President for his approval.

It seems clear that some machinery is essential now to provide the way for continuous and effective controls and to insure or provide for the necessary and persistent research and development of the possibilities of atomic energy in which the United States now leads the way. If properly controlled by the peace loving nations of the world this energy should insure the peace of the world for generations. If misused it may lead to the complete destruction of civilization.

In the circumstances I suggest that a committee of six or seven be set up at once to study and report on the whole problem of temporary war controls and publicity, and to survey and make recommendation on postwar research, development and controls, and the legislation necessary to effectuate them.

The members of this committee should be appointed by you as Secretary of War subject to the approval of the President. When appointed the committee will need promptly to organize appropriate panels to aid in its work—panels of specially qualified scientists, Army and Navy personnel, Congressional advisers, legislative draftsmen and others.

George L. Harrison

---

15. Interim Committee Minutes, May 31, 1945

10:00 A.M. to 1:15 P.M–2:25 P.M. to 4:15 P.M.

Present:

Members of the Committee
Secretary Henry L. Stimson, Chairman
Hon. Ralph A. Bard
Dr. Vannevar Bush
Hon. James F. Byrnes
Hon. William L. Clayton
Dr. Karl T. Compton

This document is found in Manhattan Engineer District-Top Secret, Harrison-Bundy files, folder 100, National Archives, Washington, D.C.

Dr. James B. Conant
Mr. George L. Harrison

Invited Scientists
Dr. J. Robert Oppenheimer
Dr. Enrico Fermi
Dr. Arthur H. Compton
Dr. E. O. Lawrence

By Invitation
General George C. Marshall
Major Gen. Leslie R. Groves
Mr. Harvey H. Bundy
Mr. Arthur Page

I. Opening Statement of the Chairman:

Secretary Stimson explained that the Interim Committee had been appointed by him, with the approval of the President, to make recommendations on temporary war-time controls, public announcement, legislation and post-war organization. The Secretary gave high praise to the brilliant and effective assistance rendered to the project by the scientists present for their great contributions to the work and their willingness to advise on the many complex problems that the Interim Committee had to face. He expressed the hope that the scientists would feel completely free to express their views on any phase of the subject.

The Committee had been termed an "Interim Committee" because it was expected that when the project became more widely known a permanent organization established by Congressional action or by treaty arrangements would be necessary.

The Secretary explained that General Marshall shared responsibility with him for making recommendations to the President on this project with particular reference to its military aspects; therefore, it was considered highly desirable that General Marshall be present at this meeting to secure at first hand the views of the scientists.

The Secretary expressed the view, a view shared by General Marshall, that this project should not be considered simply in terms of military weapons, but as a new relationship of man to the universe. This discovery might be compared to the discoveries of the Copernican theory and of the laws of gravity, but far more important than these in its effect on the lives of men. While the advances in the field to date had been fostered by the needs of war, it was important to realize that the implications of the project went far beyond the needs of the present war. It must be controlled if possible to make it an assurance of future peace rather than a menace to civilization.

The Secretary suggested that he hoped to have the following questions discussed during the course of the meeting:

1. Future military weapons.
2. Future international competition.
3. Future research.
4. Future controls.
5. Future developments, particularly non-military.

II. Stages of Development:

As a technical background for the discussions, Dr. A. H. Compton explained the various stages of development. The first stage involved the separation of uranium 235. The second stage involved the use of "breeder" piles to produce enriched materials from which plutonium or new types of uranium could be obtained. The first stage was being used to produce material for the present bomb while the second stage would produce atomic bombs with a tremendous increase in explosive power over those now in production. Production of enriched materials was now on the order or pounds or hundreds of pounds and it was contemplated that the scale of operations could be expanded sufficiently to produce many tons. While bombs produced from the products of the second stage had not yet been proven in actual operation, such bombs were considered a scientific certainty. It was estimated that from January 1946 it would take one and one-half years to prove [sic] this second stage in view of certain technical and metallurgical difficulties; that it would take three years to get plutonium in volume, and that it would take perhaps six years for any competitor to catch up with us.

Dr. Fermi estimated that approximately twenty pounds of enriched material would be needed to carry on research in current engineering problems and that a supply of one-half to one ton would be needed for research on the second stage.

In response to the Secretary's question, Dr. A. H. Compton stated that the second stage was dependent upon vigorous exploitation of the first stage and would in no way vitiate the expenditure already made on the present plant.

Dr. Conant mentioned a so-called "third stage" of development in which the products of the "second stage" would be used simply as a detonator for heavy water. He asked Dr. Oppenheimer for an estimate of the time factor involved in developing this phase. *Dr. Oppenheimer* stated that this was a far more difficult development than the previous stages and estimated that a minimum of three years would be required to reach production. He pointed out that heavy water (hydrogen) was much cheaper to produce than the other materials and could eventually be obtained in far greater quantity.

Dr. Oppenheimer reviewed the scale of explosive force involved in these several stages. One bomb produced in the first stage was estimated to have the explosive force of 2,000–20,000 tons of TNT. The actual blast effect would be accurately measured when the test was made. In the second stage the explosive force was estimated to be equal to 50,000–100,000 tons of TNT. It was considered possible that a bomb developed from the third stage might produce an explosive force equal to 10,000,000– 100,000,000 tons of TNT. . . .

VI. Russia:

In considering the problem of controls and international collaboration the question of paramount concern was the attitude of Russia. *Dr. Oppenheimer* pointed out that Russia had always been very friendly to science and suggested that we might open up this subject with them in a tentative fashion and in the most general terms without giving them any details of our productive effort. He thought we might say that a great national effort had been put into this project and express a hope for cooperation with them in this field. He felt strongly that we should not prejudge the Russian attitude in this matter.

At this point *General Marshall* discussed at some length the story of charges and counter-charges that have been typical of our relations with the Russians, pointing out that most of these allegations have proven unfounded. The seemingly uncooperative attitude of Russia in military matters stemmed from the necessity of maintaining security. He said that he had accepted this reason for their attitude in his dealings with the Russians and had acted accordingly. As to the post-war situation and in matters other than purely military, he felt that he was in no position to express a view. With regard to this field he was inclined to favor the building up of a combination among like-minded powers, thereby forcing Russia to fall in line by the very force of this coalition. General Marshall was certain that we need have no fear that the Russians, if they had knowledge of our project, would disclose this information to the Japanese. He raised the question whether it might be desirable to invite two prominent Russian scientists to witness the test.

Mr. Byrnes expressed a fear that if information were given to the Russians, even in general terms, Stalin would ask to be brought into the partnership. He felt this to be particularly likely in view of our commitments and pledges of cooperation with the British. In this connection *Dr. Bush* pointed out that even the British do not have any of our blue prints on plants. *Mr. Byrnes* expressed the view, *which was generally agreed to by all present*, that the most desirable program would be to push ahead as fast as possible in production and research to make certain that we stay ahead and at the same time make every effort to better our political relations with Russia. . . .

VIII. Effect of the Bombing on the Japanese and Their Will to Fight:

It was pointed out that one atomic bomb on an arsenal would not be much different from the effect caused by any Air Corps strike of present dimensions. However, *Dr. Oppenheimer* stated that the visual effect of an atomic bombing would be tremendous. It would be accompanied by a brilliant luminescence which would rise to a height of 10,000 to 20,000 feet. The neutron effect of the explosion would be dangerous to life for a radius of at least two-thirds of a mile.

After much discussion concerning various types of targets and the effects to be produced, *the Secretary expressed the conclusion, on which there was general agreement, that we could not give the Japanese any warning; that we could not concentrate on a civilian area; but that we should seek to make a profound psychological impression on as many of the inhabitants as possible. At the suggestion of Dr. Conant the Secretary agreed that the most desirable target would be a vital war plant employing a large number of workers and closely surrounded by workers' houses.*

There was some discussion of the desirability of attempting several strikes at the same time. *Dr. Oppenheimer's* judgment was that several strikes would be feasible. *General Groves*, however, expressed doubt about this proposal and pointed out the following objections: (1) We would lose the advantage of gaining additional knowledge concerning the weapon at each successive bombing; (2) such a program would require a rush job on the part of those assembling the bombs and might, therefore, be ineffective; (3) the effect would not be sufficiently distinct from our regular Air Force bombing program.

16. Science Panel: Recommendations on the Immediate Use of Nuclear Weapons, June 16, 1945

TOP SECRET
RECOMMENDATIONS ON THE IMMEDIATE USE OF NUCLEAR WEAPONS
June 16, 1945

You have asked us to comment on the initial use of the new weapon. This use, in our opinion, should be such as to promote a satisfactory adjustment of our international relations. At the same time, we recognize our obligation to our nation to use the weapons to help save American lives in the Japanese war.

(1) To accomplish these ends we recommend that before the weapons are used not only Britain, but also Russia, France, and China be advised that we have made considerable progress in our work on atomic weapons, that these may be ready to use during the present war, and that we would welcome suggestions as to how we can cooperate in making this development contribute to improved international relations.

(2) The opinions of our scientific colleagues on the initial use of these weapons are not unanimous: they range from the proposal of a purely technical demonstration to that of the military application best designed to induce surrender. Those who advocate a purely technical demonstration would wish to outlaw the use of atomic weapons, and have feared that if we use the weapons now our position in future negotiations will be prejudiced. Others emphasize the opportunity of saving American lives by immediate military use, and believe that such use will improve the international pros-

This text is found in Manhattan Engineer District-Top Secret, Harrison-Bundy files, folder 76, National Archives, Washington, D.C.

pects, in that they are more concerned with the prevention of war than with the elimination of this specific weapon. We find ourselves closer to these latter views; we can propose no technical demonstration likely to bring an end to the war; we see no acceptable alternative to direct military use.

(3) With regard to these general aspects of the use of atomic energy, it is clear that we, as scientific men, have no proprietary rights. It is true that we are among the few citizens who have had occasion to give thoughtful consideration to these problems during the past few years. We have, however, no claim to special competence in solving the political, social, and military problems which are presented by the advent of atomic power.

A. H. Compton
E. O. Lawrence
J. R. Oppenheimer
E. Fermi
[signed]   J. R. Oppenheimer
For the Panel

17. Stimson and the Scientists' Concerns, June 26, 1945

WAR DEPARTMENT
WASHINGTON

26 June 1945
TOP SECRET

Memorandum for the Secretary of War:

Many of the scientists who have been working on S-1 have expressed considerable concern about the future dangers of the development of atomic power. Some are fearful that no safe system of international control can be established. They, therefore, envisage the possibility of an armament race that may threaten civilization.

One group of scientists, working in the Chicago Laboratories, urges that

This text is found in Manhattan Engineer District-Top Secret, Harrison-Bundy files, folder 77, National Archives, Washington, D.C.

we should not make use of the bomb, so nearly completed, against any enemy country at this time. They feel that to do so might sacrifice our whole moral position and thus make it more difficult for us to be the leaders in proposing or enforcing any system of international control designed to make this tremendous force an influence towards the maintenance of world peace rather than an uncontrollable weapon of war.

This anonymous statement of the Chicago scientists was submitted for comment to the Panel of Scientists appointed by the Interim Committee. Their answer was that they saw no acceptable alternative to direct military use since they believe that such use would be an obvious means of saving American lives and shortening the war.

It is interesting that practically all of the scientists, including those on the panel, feel great concern for the future if atomic power is not controlled through some effective international mechanism. Accordingly, most of them believe that one of the effective steps in establishing such a control is the assurance that, after this war is over, there shall be a free interchange of scientific opinion throughout the world supplemented, if possible, by some system of inspection. This they admit is a problem of the future. In the meantime, however, they feel that we must, even before actual use, briefly advise the Russians of our progress.

This matter of notice to the Russians was made a subject of thorough discussion at the last meeting of the Interim Committee on June 21. It was unanimously agreed that in view of the importance of securing an effective future control, and in view of the fact that most of the story, other than production secrets, will become known in ———[sic] in any event, there would be considerable advantage, if a suitable opportunity arises at the "Big Three" meeting, in having the President advise the Russians simply that we are working intensely on this weapon and that, if we succeed as we think we will, we plan to use it against the enemy. Such a statement might well be supplemented by the statement that in the future, after the war, we would expect to discuss the matter further with a view to insuring that this means of warfare will become a substantial aid in preserving the peace of the world rather than a weapon of terror and destruction.

It was felt by the Committee that if the Russians should ask for more details now rather than later or if they should raise questions as to timetables, methods of production, etc., they should be told that we are not yet ready to discuss the subject beyond the simple statement suggested above. Our purpose is merely to let them know that we did not wish to proceed with actual use without giving them prior information that we intend to do so. Not to give them this prior information at the time of the "Big Three" Conference and within a few weeks thereafter to use the weapon and to make fairly complete statements to the world about its history and development, might well make it impossible ever to enlist Russian cooperation in the set-up of future international controls over this new power.

It was agreed by the Committee that in view of the provisions of the Quebec Agreement it would be desirable to discuss this whole aspect of the question with the Prime Minister in advance of the "Big Three" Conference.

[signed] George L. Harrison

18. Chicago Scientists' Petition to the President, July 17, 1945

(Drafted by Leo Szilard and signed by 68 members of the Metallurgical Laboratory in Chicago)

Discoveries of which the people of the United States are not aware may affect the welfare of this nation in the near future. The liberation of atomic power which has been achieved places atomic bombs in the hands of the Army. It places in your hands, as Commander-in-Chief, the fateful decision whether or not to sanction the use of such bombs in the present phase of the war against Japan.

We, the undersigned scientists, have been working in the field of atomic power. Until recently we have had to fear that the United States might be attacked by atomic bombs during this war and that her only defense might lie in a counterattack by the same means. Today, with the defeat of Germany, this danger is averted and we feel impelled to say what follows:

The war has to be brought speedily to a successful conclusion and attacks by atomic bombs may very well be an effective method of warfare. We feel, however, that such attacks on Japan could not be justified, at least not until the terms which will be imposed after the war on Japan were made public in detail and Japan were given an opportunity to surrender.

If such, public announcement gave assurance to the Japanese that they could look forward to a life devoted to peaceful pursuits in their homeland and if Japan still refused to surrender, our action might then, in certain

This document appeared in S. R. Weart and Gertrud Weiss, eds., *Leo Szilard: His Version of the Facts*, vol. 2 (Cambridge, Mass.: MIT Press, 1978), pp. 211–12.

circumstances, find itself forced to resort to the use of atomic bombs. Such a step, however, ought not to be made at any time without seriously considering the moral responsibilities which are involved.

The development of atomic power will provide the nations with new means of destruction. The atomic bombs at our disposal represent only the first step in this direction, and there is almost no limit to the destructive power which will become available in the course of their future development. Thus a nation which sets the precedent of using these newly liberated forces of nature for purposes of destruction may have to bear the responsibility of opening the door to an era of devastation on an unimaginable scale.

If after the war a situation is allowed to develop in the world which permits rival powers to be in uncontrolled possession of these new means of destruction, the cities of the United States as well as the cities of other nations will be in continuous danger of sudden annihilation. All the resources of the United States, moral and material, may have to be mobilized to prevent the advent of such a world situation. Its prevention is at present the solemn responsibility of the United States—singled out by virtue of her lead in the field of atomic power.

The added material strength which this lead gives to the United States brings with it the obligation of restraint and if we were to violate this obligation our moral position would be weakened in the eyes of the world and in our own eyes. It would then be more difficult for us to live up to our responsibility of bringing the unloosened forces of destruction under control.

In view of the foregoing, we, the undersigned, respectfully petition: first, that you exercise your power as Commander-in-Chief to rule that the United States shall not resort to the use of atomic bombs in this war unless the terms which will be imposed upon Japan have been made public in detail and Japan knowing these terms has refused to surrender; second, that in such an event the question whether or not to use atomic bombs be decided by you in the light of the consideration presented in this petition as well as all the other moral responsibilities which are involved.

19. White House Press Release on Hiroshima, August 6, 1945

Statement by the President of the United States[1]

Sixteen hours ago an American airplane dropped one bomb on Hiroshima, an important Japanese Army base. That bomb had more power than 20,000 tons of T.N.T. It had more than two thousand times the blast power of the British "Grand Slam" which is the largest bomb ever yet used in the history of warfare.

The Japanese began the war from the air at Pearl Harbor. They have been repaid many fold. And the end is not yet. With this bomb we have now added a new and revolutionary increase in destruction to supplement the growing power of our armed forces. In their present form these bombs are now in production and even more powerful forms are in development.

It is an atomic bomb. It is a harnessing of the basic power of the universe. The force from which the sun draws its power has been loosed against those who brought war to the Far East.

Before 1939, it was the accepted belief of scientists that it was theoretically possible to release atomic energy. But no one knew any practical method of doing it. By 1942, however, we knew that the Germans were working feverishly to find a way to add atomic energy to the other engines of war with which they hoped to enslave the world. But they failed. We may be grateful to Providence that the Germans got the V-1's and the V-2's late and in limited quantities and even more grateful that they did not get the atomic bomb at all.

The battle of the laboratories held fateful risks for us as well as the battles of the air, land and sea, and we have now won the battle of the laboratories as we have won the other battles.

Beginning in 1940, before Pearl Harbor, scientific knowledge useful in war was pooled between the United States and Great Britain, and many priceless helps to our victories have come from that arrangement. Under that general policy the research on the atomic bomb was begun. With American and British scientists working together we entered the race of discovery against the Germans.

The United States had available the large number of scientists of distinction in the many needed areas of knowledge. It had the tremendous

This text appeared in Foreign Relations of the United States, *Potsdam*, vol. 2 (Washington, D.C., 1960), pp. 1380–81.
1. Released August 6, 1945. A draft of this statement had been discussed with the British.

industrial and financial resources necessary for the project and they could be devoted to it without undue impairment of other vital war work. In the United States the laboratory work and the production plants, on which a substantial start had already been made, would be out of reach of enemy bombing, while at that time Britain was exposed to constant air attack and was still threatened with the possibility of invasion. For these reasons Prime Minister Churchill and President Roosevelt agreed that it was wise to carry on the project here. We now have two great plants and many lesser works devoted to the production of atomic power. Employment during peak construction numbered 125,000 and over 65,000 individuals are even now engaged in operating the plants. Many have worked there for two and a half years. Few know what they have been producing. They see great quantities of material going in and they see nothing coming out of these plants, for the physical size of the explosive charge is exceedingly small. We have spent two billion dollars on the greatest scientific gamble in history—we won.

But the greatest marvel is not the size of the enterprise, its secrecy, nor its cost, but the achievement of scientific brains in putting together infinitely complex pieces of knowledge held by many men in different fields of science into a workable plan. And hardly less marvelous has been the capacity of industry to design, and of labor to operate, the machines and methods to do things never done before so that the brain child of many minds came forth in physical shape and performed as it was supposed to do. Both science and industry worked under the direction of the United States Army, which achieved a unique success in managing so diverse a problem in the advancement of knowledge in an amazingly short time. It is doubtful if such another combination could be got together in the world. What has been done is the greatest achievement of organized science in history. It was done under high pressure and without failure.

We are now prepared to obliterate more rapidly and completely every productive enterprise the Japanese have above ground in any city. We shall destroy their docks, their factories, and their communications. Let there be no mistake; we shall completely destroy Japan's power to make war.

It was to spare the Japanese people from utter destruction that the ultimatum of July 26 was issued at Potsdam. Their leaders promptly rejected that ultimatum. If they do not now accept our terms they may expect a rain of ruin from the air, the like of which has never been seen on this earth. Behind this air attack will follow sea and land forces in such numbers and power as they have not yet seen and with the fighting skill of which they are already well aware.

The Secretary of War, who has kept in personal touch with all phases of this project, will immediately make public a statement giving further details.[2]

His statement will give facts concerning the sites of Oak Ridge near Knoxville, Tennessee, and at Richland near Pasco, Washington, and an installation near Santa Fe, New Mexico. Although the workers at the sites have been making materials to be used in producing the greatest destructive force in history they have not themselves been in danger beyond that of many other occupations, for the utmost care has been taken of their safety.

The fact that we can release atomic energy ushers in a new era in man's understanding of nature's forces. Atomic energy may in the future supplement the power that now comes from coal, oil, and falling water, but at present it cannot be produced on a basis to compete with them commercially. Before that comes there must be a long period of intensive research.

It has never been the habit of the scientists of this country or the policy of this Government to withhold from the world scientific knowledge. Normally, therefore, everything about the work with atomic energy would be made public.

But under present circumstances it is not intended to divulge the technical processes of production or all the military applications, pending further examination of possible methods of protecting us and the rest of the world from the danger of sudden destruction.

I shall recommend that the Congress of the United States consider promptly the establishment of an appropriate commission to control the production and use of atomic power within the United States. I shall give further consideration and make further recommendations to the Congress as to how atomic power can become a powerful and forceful influence towards the maintenance of world peace.

2. For the text of Stimson's statement of August 6, 1945, see Raymond Dennett and Robert K. Turner, eds., *Documents on American Foreign Relations, July 1, 1945–December 31, 1946* (Princeton: Princeton University Press, 1948), p. 413.

III

Atomic Energy in a Postwar World

*A*t the end of World War II, the United States had spent over two billion dollars to develop an enormous atomic industry operated by the U.S. Army. The Manhattan Engineer District had nurtured a new science and produced a weapon that won the war and stunned the world. Yet, in 1945, scientists associated with the Manhattan Project also saw a potential for the peaceful atom.

The military and scientific establishments fought for control of atomic energy after the war. The military proposed an Atomic Energy Commission within the Department of Defense that would take charge of nuclear energy. The May-Johnson Bill introduced in Congress in 1946 represented the military view that weapons were of the highest importance and that the secret of the bomb could be preserved only through military control.

Most scientists disagreed. The Federation of Atomic Scientists, made up of many of those who had worked on the bomb, lobbied in Washington for civilian control. Working privately in small groups, prominent scientists visited individual congressmen. During the war, military control had restricted exchange of information and actually hindered progress on the bomb, they said. More importantly, the scientists persuasively argued that civilian decision makers, not the military, should control nuclear weapons. As a safety measure, they wanted the agency in charge of atomic weapons to be independent of the armed forces.

The scientists won the legislative battle. The Atomic Energy Act of 1946, sponsored by Senator Brien McMahon of Connecticut, established an independent, civilian-controlled Atomic Energy Commission. A Military Liaison Committee did oversee the military activities of the Atomic Energy Commission, but the five civilian commissioners retained complete responsibility for weapons, which could not be delivered to the armed forces without a direct presidential order.

Civilian control at home gave no assurances for what happened in the international arena. Those who had witnessed the horrible destructive power of the bomb at Alamogordo or had seen photographs of Hiroshima and Nagasaki sought to ban the bomb altogether. In the spring of 1946, the United States announced its plan for the international control of atomic energy, known as the Acheson-Lilienthal Report. Its sponsors were Dean Acheson, the undersecretary of state, and David Lilienthal, former head of the Tennessee Valley Authority who would soon be named chairman of the Atomic Energy Commission. The proposal suggested a series of steps by which the nations of the world would surrender their possession of atomic weapons. During the transition, the United States would maintain its monopoly and all nations would be subject to international inspection.

In June 1946, shortly after Secretary of State George C. Marshall outlined his plan for the economic recovery of Europe, Bernard M. Baruch, the U.S. delegate to the United Nations Atomic Energy Commission, presented another plan for international control, focusing on an International Atomic Development Authority. The United States would agree to place its atomic bomb secrets with the authority provided that the great powers could establish a suitable international control and inspection system. In addition, Baruch offered to cease manufacturing weapons and destroy all existing bombs, indeed, to eliminate war altogether. The plan, Baruch announced, was "the last, best hope of earth."

The sticking point for the Soviet Union was Baruch's insistence that the Russians could not block the plan through a veto in the United Nations Security Council, especially on the issue of vote punishment for offensive nations. The Soviet Union did not want a majority vote of the Security Council to determine its atomic future. But the United States recognized that it could not maintain its nuclear monopoly for long, and only a gradual reduction in its nuclear arsenal would maintain the balance of power. Many felt that the ways to eliminate nuclear weapons had to be explored.

The positions taken by the United States and the Soviet Union did nothing to ease suspicions and mutual distrust. The opportunity for international control of atomic weapons at their birth was gone, frozen for over a decade by the icy storms of the Cold War. The collapse of negotiations at the United Nations a year after the introduction of the Baruch Plan and President Harry S. Truman's failure to achieve any disarmament agreement with the Soviet Union had driven the issue out of the international forum until 1952. Early that year, the General Assembly established a new disarmament commission, but, as before, disagreements over disclosure of weapons and an effective inspection system doomed its efforts.

Shortly after Eisenhower took office in January 1953, he received a report from an advisory group headed by Dr. J. Robert Oppenheimer. Truman's secretary of state, Dean Acheson, had asked Oppenheimer to reexamine the question of disarmament. The report argued that there were too many nuclear weapons. The United States and the Soviet Union were like "two scorpions in a bottle, each capable of killing the other, but only at the risk of his own life," Oppenheimer wrote. The American people had to understand the dangers of atomic weapons. What was required, the scientist explained, was "candor on the part of the officials, the representatives, of the people of their country." Operation Candor was born.

Operation Candor became a heated subject for debate within the Eisenhower administration. Some advisers supported Oppenheimer in advocating a frank discussion to increase the public's knowledge of nuclear danger. Admiral Lewis Strauss, the chairman of the Atomic Energy Commission, was horrified that too much candor would aid Soviet espionage but do little for the American public. Eisenhower settled the debate by borrowing from both sides. Alarmed that a nuclear war would leave everyone dead on both sides, the president said he didn't "want to scare the country to death. Can't we find some hope?" The death of Joseph Stalin in March and the explosion of a Soviet hydrogen device in August gave the appeal even more promise and urgency.

Over a series of discussions, Eisenhower developed a new approach. Each nation, he proposed, should give the United Nations a specified amount of fissionable material to form an atomic pool for peaceful uses, primarily nuclear power. Because the plan emerged over the breakfast table, insiders soon called the approach that replaced Operation Candor, "Operation Wheaties."

Eisenhower arranged to announce his Atoms for Peace plan before the United Nations after consulting with Prime Minister Winston Churchill and his advisers in Bermuda. On December 8, 1953, Eisenhower stood before the General Assembly. The Voice of America would carry his words around the world.

The first part of the speech suggested that a nuclear or thermonuclear holocaust demanded an easing of East-West relationships, but the heart of Eisenhower's message was an appeal for peace. The powers had to find a way "to serve the needs rather than the fears of mankind." By establishing a stockpile of uranium under an International Atomic Energy Agency, the atom might "serve the peaceful pursuits of mankind." Concluding, Eisenhower promised that the United States would "devote its entire heart and mind to find the way by which the miraculous inventiveness of man shall not be dedicated to his death, but consecrated to his life."

The audience of thirty-five hundred people sat silently until the final words, then burst into enthusiastic applause unprecedented in United Nations history. The delegations from the Soviet bloc countries joined the cheers. The Atoms for Peace proposal had offered hope to a fearful world.

Eisenhower's speech, full of idealism, did not address the thorny issues of nuclear proliferation, of nuclear disarmament, or of verifiable inspection systems. Rather, the president sought to avoid the old barrier and went his way to a simple international agreement before proceeding with controversial proposals on general disarmament. But within a year, the proposal was foundering due to mutual American and Soviet suspicions. The United States wanted to bring the Soviets to some terms before their atomic stockpile grew too large. The Russians sought further delays. Not until October 1956 was the International Atomic Energy Agency established with the participation of seventy-two nations, including the Soviet Union. Although it has allocated fissionable materials to other nations for industrial purposes, its main contribution has been the forum for the establishment of international standards for radiation protection.

Valuable memoirs on these years are Lewis L. Strauss, Men and Decisions (Garden City, N.Y.: Doubleday, 1962), and Dwight D. Eisenhower, Mandate for Change, 1953–1956: The White House Years (Garden City, N.Y.: Doubleday, 1963). See also U.S. Congress, Senate, Committee on Foreign Relations, Atoms for Peace Manual, 84 Cong., 1 sess., 1955, Senate Report 55; and more recently Herbert S. Parmet, Eisenhower and the American Crusades (New York: MacMillan, 1972).

On the postwar years, see both Richard G. Hewlett and Oscar E. Anderson, The New World, 1939–1946 (University Park, Pennsylvania State University Press, 1962), and Richard G. Hewlett and Francis Duncan, Atomic Shield, 1947–1952, in the same series (1969). On the international control of atomic energy after 1945, see Walter LaFeber, America, Russia, and the Cold War, 1945–1966 (New York: Wiley, 1967) and the United Nations Department of Political and Security Council Affairs's The United Nations and Disarmament, 1945–1970 (New York: United Nations, 1970). On the postwar role of the atomic bomb in American foreign policy, see Gregg Herken, The Winning Weapon: The Atomic Bomb in the Cold War, 1945–1950 (New York: Alfred A. Knopf, 1980).

20. Henry Stimson to Harry S. Truman, September 11, 1945

Dear Mr. President:

In handing you today my memorandum about our relations with Russia in respect to the atomic bomb, I am not unmindful of the fact that when in Potsdam I talked with you about the question whether we could be safe in sharing the atomic bomb with Russia while she was still a police state and before she put into effect provisions assuring personal rights of liberty to the individual citizen.

I still recognize the difficulty and am still convinced of the ultimate importance of a change in Russian attitude toward individual liberty but I have come to the conclusion that it would not be possible to use our possession of the atomic bomb as a direct lever to produce the change. I have become convinced that any demand by us for an internal change in Russia as a condition of sharing in the atomic weapon would be so resented that it would make the objective we have in view less probable.

I believe that the change in attitude toward the individual in Russia will come slowly and gradually and I am satisfied that we should not delay our approach to Russia in the matter of the atomic bomb until that process has been completed. My reasons are set forth in the memorandum I am handing you today. Furthermore, I believe that this long process of change in Russia is more likely to be expedited by the closer relationship in the matter of the atomic bomb which I suggest and the trust and confidence that I believe would be inspired by the method of approach which I have outlined.

> Faithfully yours,
> Henry L. Stimson
> Secretary of War.

The President,
The White House

MEMORANDUM FOR THE PRESIDENT

Subject: The advent of the atomic bomb has stimulated great military and probably even greater political interest throughout the civilized world. In a

This text appeared in Foreign Relations of the United States, *1945*, vol. 2 (Washington, D.C., 1971), pp. 40–44.

world atmosphere already extremely sensitive to power, the introduction of this weapon has profoundly affected political considerations in all sections of the globe.

In many quarters it has been interpreted as a substantial offset to the growth of Russian influence on the continent. We can be certain that the Soviet Government has sensed this tendency and the temptation will be strong for the Soviet political and military leaders to acquire this weapon in the shortest possible time. Britain in effect already has the status of a partner with us in the development of this weapon. Accordingly, unless the Soviets are voluntarily invited into the partnership upon a basis of cooperation and trust, we are going to maintain the Anglo-Saxon bloc over against the Soviet in the possession of this weapon. Such a condition will almost certainly stimulate feverish activity on the part of the Soviet toward the development of this bomb in what will in effect be a secret armament race of a rather desperate character. There is evidence to indicate that such activity may have already commenced.

If we feel, as I assume we must, that civilization demands that some day we shall arrive at a satisfactory international arrangement respecting the control of this new force, the question then is how long we can afford to enjoy our momentary superiority in the hope of achieving our immediate peace council objectives.

Whether Russia gets control of the necessary secrets of production in a minimum of say four years or a maximum of twenty years is not nearly as important to the world and civilization as to make sure that when they do get it they are willing and cooperative partners among the peace-loving nations of the world. It is true if we approach them now, as I would propose, we may be gambling on their good faith and risk their getting into production of bombs a little sooner than they would otherwise.

To put the matter concisely, I consider the problem of our satisfactory relations with Russia as not merely connected with but as virtually dominated by the problem of the atomic bomb. Except for the problem of the control of the bomb, those relations, while vitally important, might not be immediately pressing. The establishment of relations of mutual confidence between her and us could afford to wait the slow progress of time. But with the discovery of the bomb they became immediately emergent. Those relations may be perhaps irretrievably embittered by the way in which we approach the solution of the bomb with Russia. For if we fail to approach them now and merely continue to negotiate with them, having this weapon rather ostentatiously on our hip, their suspicions and their distrust of our purposes and motives will increase. It will inspire them to greater efforts in an all-out effort to solve the problem. If the solution is achieved in that spirit, it is much less likely that we will ever get the kind of covenant we may desperately need in the future. This risk, is, I believe, greater than the other, inasmuch as our objective must be to get the best kind of international bargain we can—one that has some chance of being kept and saving civilization not for five or for twenty years, but forever.

The chief lesson I have learned in a long life is that the only way you can make a man trustworthy is to trust him; and the surest way to make him untrustworthy is to distrust him and show your distrust.

If the atomic bomb were merely another though more devastating military weapon to be assimilated into our pattern of international relations, it would be one thing. We could then follow the old custom of secrecy and nationalistic military superiority relying on international caution to prescribe the future use of the weapon as we did with gas. But I think the bomb instead constitutes merely a first step in a new control by man over the forces of nature too revolutionary and dangerous to fit into the old concepts. I think it really caps the climax of the age between man's growing technical power for destructiveness and his psychological power of self-control and group control—his moral power. If so, our method of approach to the Russians is a question of the most vital importance in the evolution of human progress.

Since the crux of the problem is Russia, any contemplated action leading to the control of this weapon should be primarily directed *to* Russia. It is my judgment that the Soviet would be more apt to respond sincerely to a direct and forthright approach made by the United States on the subject than would be the case if the approach were made as a part of a general international scheme, or if the approach were made after a succession of express or implied threats or near threats in our peace negotiations.

My idea of an approach to the Soviets would be a direct proposal after discussion with the British that we would be prepared in effect to enter an arrangement with the Russians, the general purpose of which would be to control and limit the use of the atomic bomb as an instrument of war and so far as possible to direct and encourage the development of atomic power for peaceful and humanitarian purposes. Such an approach might more specifically lead to the proposal that we would stop work on the further improvement in, or manufacture of, the bomb as a military weapon, provided the Russians and the British would agree to do likewise. It might also provide that we would be willing to impound what bombs we now have in the United States provided the Russians and the British would agree with us that in no event will they or we use a bomb as an instrument of war unless all three Governments agree to that use. We might also consider including in the arrangement a covenant with the U. K. and the Soviets providing for the exchange of benefits of future development whereby atomic energy may be applied on a mutually satisfactory basis for commercial or humanitarian purposes.

I would make such an approach just as soon as our immediate political considerations make it appropriate.

I emphasize perhaps beyond all other considerations the importance of taking this action with Russia as a proposal of the United States—backed by Great Britain but peculiarly the proposal of the United States. Action of any international group of nations, including many small nations who have not demonstrated their potential power or responsibility in this war would not, in my opinion, be taken seriously by the Soviets. The loose debates which would surround such proposal, if put before a conference of nations, would provoke but scant favor from the Soviet. As I say, I think this is the most important point in the program.

After the nations which have won this war have agreed to it, there will be ample time to introduce France and China into the covenants and finally to incorporate the agreement into the scheme of the United Nations. The use of this bomb has been accepted by the world as the result of the initiative and productive capacity of the United States, and I think this factor is a most potent lever toward having our proposals accepted by the Soviets, whereas I am most skeptical of obtaining any tangible results by way of any international debate. I urge this method as the most realistic means of accomplishing this vitally important step in the history of the world.

Henry L. Stimson
"Secretary of War."

21. The McMahon Bill, December 20, 1945 (Atomic Energy Act of 1946)

A BILL

For the development and control of atomic energy.
Be it enacted by the Senate and House of Representatives of the United States of America in congress assembled,

DECLARATION OF POLICY

Section 1. (a) Findings and Declaration.—Research and experimentation in the field of nuclear fission have attained the stage at which the release of atomic energy on a large scale is practical. The significance of the atomic bomb for military purposes is evident. The effect of the use of atomic energy for civilian purposes upon the social, economic, and political structures of today cannot now be determined. It is reasonable to anticipate, however, that tapping this new source of energy will cause profound changes in our present way of life. Accordingly, it is hereby declared to be the policy of the people of the United States that the development and utilization of atomic energy shall be directed toward improving the public welfare, increasing the standard of living, strengthening free competition among private enterprises so far as practicable, and cementing world peace.

(b) Purpose of Act.—It is the purpose of this Act to effectuate these policies by providing, among others, for the following major programs;

(1) A program of assisting and fostering private research and development on a truly independent basis to encourage maximum scientific progress;

(2) A program for the free dissemination of basic scientific information and for maximum liberality in dissemination of related technical information;

(3) A program of federally conducted research to assure the Government of adequate scientific and technical accomplishments;

(4) A program for Government control of the production, ownership, and use of fissionable materials to protect the national security and to insure the broadest possible exploitation of the field;

This text appeared in Senate Special Committee on Atomic Energy, *Atomic Energy Act of 1946. Hearings on S.1717 January 22–April 4, 1946* (Washington, D.C., 1946), pp. 1–9.

(5) A program for simultaneous study of the social, political, and economic effects of the utilization of atomic energy; and

(6) A program of administration which will be consistent with international agreements made by the United States, and which will enable the Congress to be currently informed so as to take further legislative action as may hereafter be appropriate.

ATOMIC ENERGY COMMISSION

Sec. 2. (a) There is hereby established an Atomic Energy Commission (herein called the Commission), which shall be composed of five members. Three members shall constitute a quorum of the Commission. The President shall designate one member as a Chairman of the Commission.

(b) Members of the Commission shall be appointed by the President, by and with the advice and consent of the Senate, and shall serve at the pleasure of the President. In submitting nominations to the Senate, the President shall set forth the experience and qualifications of each person so nominated. Each member, except the Chairman, shall receive compensation at the rate of $15,000 per annum; the Chairman shall receive compensation at the rate of $20,000 per annum. No member of the Commission shall engage in any other business, vocation, or employment than that of serving as a member of the Commission.

(c) The principal office of the Commission shall be in the District of Columbia, but the Commission may exercise any or all of its powers in any place. The Commission shall hold such meetings, conduct such hearings, and receive such reports as will enable it to meet its responsibilities for carrying out the purpose of this Act.

RESEARCH

Sec. 3. (a) Research Assistance.—The Commission is directed to exercise its powers in such a manner as to insure the continued conduct of research and developmental activities in the fields specified below by private or public institutions or persons and to assist in the acquisition of an ever-expanding fund of theoretical and practical knowledge in such fields. To this end the Commission is authorized and directed to make contracts, agreements, arrangements, grants-in-aid, and loans—

(1) for the conduct of research and developmental activities relating to (a) nuclear processes; (b) the theory and production of atomic energy, including processes and devices related to such production; (c) utilization of fissionable and radioactive materials for medical or health

purposes; (d) utilization of fissionable and radioactive materials for all other purposes, including industrial uses; and (e) the protection of health during research and production activities; and

(2) for studies of the social, political, and economic effects of the availability and utilization of atomic energy.

The Commission may make partial advance payments on such contracts and arrangements. Such contracts or other arrangements may contain provisions to protect health, to minimize danger from explosion, and for reporting and inspection of work performed thereunder as the Commission may determine, but shall not contain any provisions or conditions which prevent the dissemination of scientific or technical information, except to the extent already required by the Espionage Act.

(b) Federal Atomic Research.—The Commission is authorized and directed to conduct research and developmental activities through its own facilities in the fields specified in (a) above.

PRODUCTION OF FISSIONABLE MATERIALS

Sec. 4. (a) Definition.—The term "production of fissionable materials" shall include all methods of manufacturing, producing, refining, or processing fissionable materials, including the process of separating fissionable material from other substances in which such material may be contained, whether by thermal diffusion, electromagnetic separation, or other processes.

(b) Authority to Produce.—The Commission shall be the exclusive producer of fissionable materials, except production incident to research or developmental activities subject to restrictions provided in subparagraph (d) below. The quantities of fissionable material to be produced in any quarter shall be determined by the President.

(c) Prohibition.—It shall be unlawful for any person to produce any fissionable material except as may be incident to the conduct of research or developmental activities.

(d) Research and Development on Production Processes. (1) The Commission shall establish by regulation such requirements for the reporting of research and developmental activities on the production of fissionable materials as will assure the Commission of full knowledge of all such activities, rates of production, and quantities produced.

(2) The Commission shall provide for the frequent inspection of all such activities by employees of the Commission.

(3) No person may in the course of such research or developmental

activities possess or operate facilities for the production of fissionable material in quantities or at a rate sufficient to construct a bomb or other military weapon unless all such facilities are the property of and subject to the control of the Commission. The Commission is authorized, to the extent that it deems such action consistent with the purposes of this Act, to enter into contracts for the conduct of such research or developmental activities involving the use of the Commission's facilities.

(e) Existing Contracts.—The Commission is authorized to continue in effect and modify such contracts for the production of fissionable materials as may have been made prior to the effective date of this Act, except that, as rapidly as practicable, and in any event not more than one year after the effective date of this Act, the Commission shall arrange for the exclusive operation of facilities employed in the manufacture of fissionable materials by employees of the Commission.

CONTROL OF MATERIALS

Sec. 5. (a) (1) Definition.—The term "fissionable materials" shall include plutonium, uranium 235, and such other materials as the Commission may from time to time determine to be capable of releasing substantial quantities of energy through nuclear fission of the materials.

(2) Privately Owned Fissionable Materials.—Any person owning any right, title, or interest in or to any fissionable material shall forthwith transfer all such right, title, or interest to the Commission.

(3) Prohibition.—It shall be unlawful for any person to (a) own any fissionable material; or (b) after sixty days after the effective date of this Act and except as authorized by the Commission possess any fissionable material; or (c) export from or import into the United States any fissionable material, or directly or indirectly be a party to or in any way a beneficiary of, any contract, arrangement or other activity pertaining to the production, refining, or processing of any fissionable material outside of the United States.

(4) Distribution of Fissionable Materials.—The Commission is authorized and directed to distribute fissionable materials to all applicants requesting such materials for the conduct of research or developmental activities either independently or under contract or other arrangement with the Commission. If sufficient materials are not available to meet all such requests, and applications for licenses under section 7, the Commission shall allocate fissionable materials among all such applicants in the manner best calculated to encourage independent research and development by making adequate fissionable materials available for such purposes. The Commission

shall refuse to distribute or allocate any materials to any applicant, or shall recall any materials after distribution or allocation from any applicant, who is not equipped or who fails to observe such safety standards to protect health and to minimize danger from explosion as may be established by the Commission.

(b) Source Materials.—

(1) Definition.—The term "source materials" shall include any ore containing uranium, thorium, or beryllium, and such other materials peculiarly essential to the production of fissionable materials as may be determined by the Commission with the approval of the President.

(2) License for Transfers Required.—No person may transfer possession or title to any source material after mining, extraction, or removal from its place of origin, and no person may receive any source material without a license from the Commission.

(3) Issuance of Licenses.—Any person desiring to transfer or receive possession of any source material shall apply for a license therefore in accordance with such procedures as the Commission may by regulation establish. The Commission shall establish such standards for the issuance or refusal of licenses as it may deem necessary to assure adequate source materials for production, research or developmental activities pursuant to this Act or to prevent the use of such materials in a manner inconsistent with the national welfare.

(4) Reporting.—The Commission is authorized to issue such regulations or orders requiring reports of ownership, possession, extraction, refining, shipment or other handling of source materials as it may deem necessary.

(c) Byproduct Materials.—

(1) Definition.—The term "byproduct material" shall be deemed to refer to all materials (except fissionable material) yielded in the processes of producing fissionable material.

(2) Distribution.—The Commission is authorized and directed to distribute, with or without charge, byproduct materials to all applicants seeking such materials for research or developmental work, medical therapy, industrial uses, or such other useful applications as may be developed. If sufficient materials to meet all such requests are not available, the Commission shall allocate such materials among applicants therefor, giving preference to the use of such materials in the conduct of research and developmental activity and medical therapy. The Commission shall refuse to distribute or allocate any byproduct materials to any applicant, or recall any materials after distribution or allocation from any applicant, who is not

equipped or who fails to observe such safety standards to protect health as may be established by the Commission.

(d) General Provisions.—(1) The Commission is authorized to—

(i) acquire or purchase fissionable or source materials within the United States or elsewhere;

(ii) take, requisition, or condemn within the United States any fissionable or source material and make just compensation therefor. The Commission shall determine such compensation. In the exercise of such rights of eminent domain and condemnation, proceedings may be instituted under the Act of August 1, 1888 (U. S. C. 1940, title 40, sec. 257), or any other applicable Federal statute. Upon or after the filing of the condemnation petition, immediate possession may be taken and the property may be treated by the Commission in the same manner as other similar property owned by it;

(iii) conduct exploratory operation, investigations, inspections to determine the location, extent, mode of occurrence, use, or condition of source materials with or without the consent of the owner of any interest therein, making just compensation for any damage or injury occasioned thereby.

(2) The Commission shall establish by regulation a procedure by which any person who is dissatisfied with its action in allocating, refusing to allocate or in rescinding any allocation of fissionable, source, or byproduct materials to him may obtain a review of such determination by a board of appeal consisting of two or more members appointed by the Commission and at least one member of the Commission.

MILITARY APPLICATIONS OF ATOMIC POWER

Sec. 6. (a) The Commission is authorized and directed to—

(1) conduct experiments and do research and developmental work in the military application of atomic power; and

(2) have custody of all assembled or unassembled atomic bombs, bomb parts, or other atomic military weapons, presently or hereafter produced, except that upon the express finding of the President that such action is required in the interests of national defense, the Commission shall deliver such quantities of weapons to the armed forces as the President may specify.

(b) The Commission shall not conduct any research or developmental work in the military application of atomic power if such research or developmental work is contrary to any international agreement of the United States.

(c) The Commission is authorized to engage in the production of atomic bombs, bomb parts, or other applications of atomic power as military weapons, only to the extent that the express consent and direction of the President of the United States has been obtained, which consent and direction shall be obtained for each quarter.

(d) It shall be unlawful for any person to manufacture, produce, or process any device or equipment designed to utilize fissionable materials as a military weapon, except as authorized by the Commission.

ATOMIC ENERGY DEVICES

Sec. 7. (a) License Required.—It shall be unlawful for any person to operate any equipment or device utilizing fissionable materials without a license issued by the Commission authorizing such operation.

Issuance of Licenses.—Any person desiring to utilize fissionable materials in any such device or equipment shall apply for a license therefor in accordance with such procedures as the Commission may by regulation establish. The Commission is authorized and directed to issue such a license on a nonexclusive basis and to supply appropriate quantities of fissionable materials to the extent available to any applicant (1) who is equipped to observe such safety standards to protect health and to minimize danger from explosion as the Commission may establish; and (2) who agrees to make available to the Commission such technical information and data concerning the operation of such device as the Commission may determine necessary to encourage the use of such devices by as many licensees as possible. Where any license might serve to maintain or foster the growth of monopoly, restraint of trade, unlawful competition, or other trade position inimical to the entry of new, freely competitive enterprises, the Commission is authorized and directed to refuse to issue such license or to establish such conditions to prevent these results as the Commission, in consultation with the Attorney General, may determine. The Commission shall report promptly to the Attorney General any information it may have of the use of such devices which appears to have these results. No license may be given to a foreign government or to any person who is not under and within the jurisdiction of the United States.

(c) Byproduct Power.—If in the production of fissionable materials the production processes yield energy capable of utilization, such energy may be used by the Commission, transferred to other Government agencies, sold to public or private utilities under contract providing for reasonable resale prices, or sold to private consumers at reasonable rates and on as broad a basis of eligibility as the Commission may determine to be possible.

(d) Reports to Congress.—Whenever in its opinion industrial, commercial, or other uses of fissionable materials have been sufficiently developed to be of practical value, the Commission shall prepare a report to the Congress stating all the facts, the Commission's estimate of the social, political, and economic effects of such utilization, and the Commission's recommendations for necessary or desirable supplemental legislation. Until such a report has been filed with the Commission and the period of ninety days has elapsed after such filing within which period the Commission may adopt supplemental legislation, no license for the use of atomic energy devices shall be issued by the Commission.

PROPERTY OF THE COMMISSION

Sec. 8. (a) The President shall direct the transfer to the Commission of the following property owned by the United States or any of its agencies, or any interest in such property held in trust for or on behalf of the United States:

(1) All fissionable materials; all bombs and bomb parts; all plants, facilities, equipment, and materials for the processing or production of fissionable materials, bombs, and bomb parts; all processes and technical information of any kind, and the source thereof (including data, drawings, specifications, patents, patent applications, and other sources, relating to the refining or production of fissionable materials; and all contracts, agreements, leases, patents, applications for patents, inventions and discoveries (whether patented or unpatented), and other rights of any kind concerning any such items;

(2) All facilities and equipment, and materials therein, devoted primarily to atomic energy research and development; and

(3) All property in the custody and control of the Manhattan engineer district.

(b) In order to render financial assistance to those States and local governments in which the activities of the Commission are carried on and in which the Commission, or its agents, have acquired properties previously subject to State and local taxation, the Commission is authorized to make payments to State and local governments in lieu of such taxes. Such payments may be in the amounts, at the times, and upon the terms the Commission deems appropriate, but the Commission shall be guided by the policy of not exceeding the taxes which would have been payable for such property in the condition in which it was acquired, except where special burdens have been cast upon the State or local government by activities of the Commission, the Manhattan engineer district, or their agents, and in such cases any benefits accruing to the States and local governments by reason of these activities shall be considered in the determination of such payments. The Commission and any corporation created by it, and the property and income of the Commission or of such corporation, are hereby expressly exempted from taxation in any manner or form by any State, county, municipality, or any subdivision thereof.

DISSEMINATION OF INFORMATION

Sec. 9. (a) Basic Scientific Information.—Basic scientific information in the fields specified in section 3 may be freely disseminated. The term "basic scientific information" shall include, in addition to theoretical knowledge of nuclear and other physics, chemistry, biology, and therapy, all results capable of accomplishment, as distinguished from the processes or techniques of accomplishing them.

(b) Related Technical Information.—The Commission shall establish a Board of Commission. The Board shall, under the direction and supervision of the Commission, provide for the dissemination of related technical information with the utmost liberality as freely as may be consistent with the foreign and domestic policies established by the President and shall have authority to—

(1) establish such information services, publications, libraries, and other registers of available information as may be helpful in effectuating this policy;

(2) designate by regulation the types of related technical information the dissemination of which will effectuate the foregoing policy. Such designations shall constitute an administrative determination that such information is not of value to the national defense and that any person is entitled to receive such information, within the meaning of the Espionage Act. Failure to make any such designation shall not, however, be deemed a determination that such undesignated information is subject to the provisions of said Act;

(3) by regulation or order, require reports of the conduct of independent research or development activities in the fields specified in section 3 and of the operation of atomic energy devices under licenses issued pursuant to section 7;

(4) provide for such inspections of independent research and development activities of the types specified in section 3 and of the oper-

ation of atomic energy devices as the Commission or the Board may determine; and

(5) whenever it will facilitate the carrying out of the purposes of the Act, adopt by regulation administrative interpretations of the Espionage Act except that any such interpretation shall, before adoption, receive the express approval of the President.

PATENTS

Sec. 10. (a) Whenever any person invents a device or method for the production, refining, or processing of fissionable material: (i) he may file a patent application to cover such invention, sending a copy thereof to the Commission; (ii) if the Commissioner of Patents determines that the invention is patentable, he shall issue a patent in the name of the Commission; and (iii) the Commission shall make just compensation to such person. The Commission shall appoint a Patent Royalty Board consisting of one or more employees and at least one member of the Commission, and the Commissioner of Patents. The Patent Royalty Board shall determine what constitutes just compensation in each such case and whether such compensation is to be paid in periodic payments rather than in a lump sum. Any person to whom any such patent has heretofore been issued shall forthwith transfer all right, title, and interest in and to such patent to the Commission and shall receive therefor just compensation as provided above.

(b) (1) Any patent now or hereafter issued covering any process or device utilizing or peculiarly necessary to the utilization of fissionable materials, or peculiarly necessary to the conduct of research or developmental activities in the fields specified in section 3, is hereby declared to be affected with the public interest and its general availability for such uses is declared to be necessary to affectuate the purposes of this Act.

(2) Any person to whom any such patent has been issued, or any person desiring to use any device or process covered by such patent for such uses, may apply to the Patent Royalty Board, for determination by such Board of a reasonable royalty fee for such use of the patented process or device intended to be used under the Commission's license.

(3) In determining such reasonable royalty fee, the Patent Royalty Board shall take into consideration any defense, general or special, that might be pleaded by a defendant in an action for infringement, the extent to which, if any, such patent was developed through federally financed research, the degree of utility, novelty, and importance of the patent, the cost to the patentee of developing such process or device, and a reasonable rate of return on such research investment by the patentee.

(4) No court, Federal, State, or Territorial, shall have jurisdiction or power to stay, restrain, or otherwise enjoin any such use of any such patented device or process by any person on the ground of infringement of such patent. In any action for infringement of any such patent filed in any such court, the court shall have authority only to order the payment of reasonable royalty fees and attorney's fees and court costs as damages for any such infringement. If the Patent Royalty Board has not previously determined the reasonable royalty fee for the use of the patented device or process involved in any case, the court in such case shall, before entering judgment, obtain from the Patent Royalty Board a report containing its recommendation as to the reasonable royalty fee it would have established had application been made to it as provided in subparagraphs 2 and 3 above.

ORGANIZATIONAL AND GENERAL AUTHORITY

Sec. 11. (a) Organization.—There are hereby established within the Commission a Division of Research, a Division of Production, a Division of Materials, and a Division of Military Application. Each division shall be under the direction of a Directory who shall be appointed by the President, by and with the advice and consent of the Senate, and shall receive compensation at the rate of $15,000 per annum. The Commission shall delegate to each such division such of its powers under this Act as in its opinion from time to time will promote the effectuation of the purposes of this Act in an efficient manner. Nothing in this paragraph shall prevent the Commission from establishing such additional divisions or other subordinate organizations as it may deem desirable.

(b) General Authority.—In the performance of its functions the Commission is authorized to—

(1) establish advisory boards to advise with and make recommendations to the Commission on legislation, policies, administration, and research;

(2) establish by regulation or order such standards and instructions to govern the possession and use of fissionable and byproduct materials as the Commission may deem necessary or desirable to protect health or to minimize danger from explosion;

(3) make such studies and investigations, obtain such information, and hold such hearings as the Commission may deem necessary or proper to assist it in exercising any authority provided in this Act, or in the administration or enforcement of this Act, or any regulations or

orders issued thereunder. For such purposes the Commission is authorized to require any person to permit the inspection and copying of any records or other documents, to administer oaths and affirmations, and by subpena to require any person to appear and testify, or to appear and produce documents, or both, at any designated place. Witnesses subpenaed under this subsection shall be paid the same fees and mileage as are paid witnesses in the district courts of the United States;

(4) create or organize corporations, the stock of which shall be wholly owned by the United States and controlled by the Commission, to carry out the provisions of this Act;

(5) appoint and fix the compensation of such officers and employees as may be necessary to carry out the functions of the Commission. All such officers and employees shall be appointed in accordance with the civil-service laws and their compensation fixed in accordance with the Classification Act of 1923, as amended, except that expert administrative, technical, and professional personnel may be employed and their compensation fixed without regard to such laws. The Commission shall make adequate provision for administrative review by a board consisting of one or more employees and at least one member of the Commission of any determination to dismiss any scientific or professional employee; and

(6) acquire such materials, property, equipment, and facilities, establish or construct such buildings and facilities, modify such building and facilities from time to time, and construct, acquire, provide, or arrange for such facilities and services for the housing, health, safety, welfare, and recreation of personnel employed by the Commission as it may deem necessary.

ENFORCEMENT

Sec. 12. (a) Any person who willfully violates, attempts to violate, or conspires to violate, any of the provisions of this Act or any regulations or orders issued thereunder shall, upon conviction thereof, be punishable by a fine of not more than $10,000, or by imprisonment for a term of not exceeding five years, or both.

(b) Whenever in the judgment of the commission any person has engaged or is about to engage in any acts or practices which constitute or will constitute a violation of any provision of this Act, it may make application to the appropriate court for an order enjoining such acts or practices, or for an order enforcing compliance with such provision, and upon a showing by the Commission that such person has engaged or is about to engage in any such acts or practices a permanent or temporary injunction, restraining order, or other order shall be granted without bond.

(c) In case of contumacy by, or refusal to obey a subpena served upon, any person pursuant to section 11 (b) (3), the district court for any district in which such person is found or resides or transacts business, upon application by the Commission, shall have jurisdiction to issue an order requiring such person to appear and give testimony or to appear and produce documents, or both; and any failure to obey such order of the court may be punished by such court as a contempt thereof.

REPORTS

Sec. 13. The Commission shall, on the first days of January, April, July, and October, submit reports to the President, to the Senate and to the House of Representatives. Such reports shall summarize and appraise the activities of the Commission and of each division and board thereof, and specifically shall contain financial statements; lists of licenses issued, of property acquired, of research contracts and arrangements entered into, and of the amounts of fissionable material and the persons to whom allocated; the Commission's program for the following quarter including lists of research contracts and arrangements proposed to be entered into; conclusions drawn from studies of the social, political, and economic effects of the release of atomic energy; and such recommendations for additional legislation as the Commission may deem necessary or desirable.

DEFINITIONS

Sec. 14. As used in this Act—

(a) The term "atomic energy" shall include all forms of energy liberated in the artificial transmutation of atomic species.

(b) The term "Government agency" means any executive department, board, bureau, commission, or other agency in the executive branch of the Federal Government, or any corporation wholly owned (either directly or through one or more corporations) by the United States.

(c) The term "person" means any individual, corporation, partnership, firm, association, trust, estate, public or private institution, group, any government other than the United States, any political subdivision of any such government, and any legal successor, representative, agent, or agency of the foregoing, or other entity.

(d) The term "United States" includes all Territories and possessions of the United States.

APPROPRIATIONS

Sec. 15. There are hereby authorized to be appropriated such sums as may be necessary and appropriate to carry out the provisions and purposes of this Act. Funds appropriated to the Commission shall, if obligated during the fiscal year for which appropriated, remain available for expenditure for four years following the expiration of the fiscal year for which appropriated. After such four-year period, the unexpended balances of appropriations shall be carried to the surplus fund and covered into the Treasury.

SEPARABILITY OF PROVISIONS

Sec. 16. If any provision of this Act, or the application of such provision to any person or circumstances, is held invalid, the remainder of this Act or the application of such provision to persons of circumstances other than those to which it is held invalid, shall not be affected thereby.

SHORT TITLE

Sec. 17. This Act may be cited as the "Atomic Energy Act of 1946."

22. The Baruch Plan

Memorandum by President Truman to the United States
Representative on the Atomic Energy Commission (Baruch)

Confidential   Washington, June 7, 1946.

Memorandum for Mr. Baruch: Because you requested it I herewith attach a statement of the United States policy with reference to atomic energy. This statement is solely for your guidance in your deliberations as the Representative of the United States on the Atomic Energy Commission of the United Nations.

The statement is general in character because I want you to have authority to exercise your judgment as to the method by which the stated objectives can be accomplished.

If as negotiations progress you conclude that there should be changes in

This text is found in Foreign Relations of the United States, 1946, vol. 1 (Washington, D.C., 1972) pp. 846–51.

this statement of policy, I will expect you to advise me and to frankly give me your views.

I know that you will keep me advised as to the negotiations. However, I want you to know that I am relying upon you to exercise your own discretion in those negotiations, subject only to the general statement of policy attached, unless you should receive from me through the Secretary of State a further statement of policy.

Harry S. Truman

(Annex)

June 7, 1946.

Statement of United States Policy

The proposals in this paper, put forth as a basis of discussion, grow out of three basic conclusions:

1. It is believed that an international agreement leaving the development of atomic energy in national hands, subject to an obligation not to develop atomic energy for war purposes and relying solely on an international inspection system to detect evasions, will not provide adequate security and indeed may be a source of insecurity.

2. It is believed

(a) That a treaty merely outlawing possession or use of the atomic bomb would not be an effective fulfillment of the directions under which the Commission is to proceed; therefore, that an international atomic development authority be set up, with adequate powers;

(b) That in connection with the greatest safeguards which can be established through a competent international authority in this field, there should be a clear statement of the consequences of violations of the system of control, including definitions of the acts which would constitute such violations and the penalties and concerted action which would follow such violations;

(c) that one of the objectives of the plan should be that when the system of control is fully in operation there would be no stockpiles of bombs in existence;

(d) That the plan might also include a parallel statement as to a system of control for biological warfare.

3. It is further believed that the aim of preventing atomic warfare can only be achieved by entrusting to an international organization

(a) Managerial control of all atomic energy activities intrinsically dangerous to world security;

(b) Power to control, inspect, and license all other activities and stages.

If an international agency is given sole responsibility for the dangerous activities, leaving the non-dangerous open to nations and their citizens and if the international agency is given and carries forward affirmative development responsibility, furthering among other things the beneficial uses of atomic energy and enabling itself to comprehend and therefore detect the misuse of atomic energy, these afford the best prospect of security.

For purposes of discussion, the following measures are proposed as representing the fundamental features of a plan which would give effect to the conclusions just stated. In this paper the proposed international agency is referred to as the Atomic Development Authority.

1. General—The Atomic Development Authority should seek to set up a thorough plan of control through various "forms of ownership, dominion, licenses, operation, inspection, research and management by competent personnel."

It is believed that the plan of control in all its aspects must be adequate not only in concept—a combination of responsibility for developments as well as control—but in type of organization and in choice of personnel to guarantee the most effective control required to provide for the security of the nations.

2. Raw Materials—the Atomic Development Authority when set up should have as one of its earliest purposes to bring under its complete dominion world supplies of uranium and thorium. The precise pattern of control for various types of deposits of such materials will have to depend upon the geological, mining, refining, and economic facts involved in different situations.

The Authority should conduct continuous surveys so that it will have the most complete knowledge of the world geology of uranium and thorium. The agency should also constantly investigate new methods for recovering these materials where they occur in small quantities so that as their recovery from such sources becomes practical, means of control can be devised.

3. Primary Production Plants—The Atomic Development Authority should exercise complete managerial control of the production of fissionable materials. This means that it should control and operate all plants producing fissionable materials in dangerous quantities and own and control the product of these plants.

4. Atomic Explosives—The Authority should be given exclusive authority to conduct research in the field of atomic explosives. Research activities in the field of atomic explosives are essential in order that the Authority may keep in the forefront of knowledge in the field of atomic energy and fulfill the objective of preventing illicit manufacture of bombs. Only by preserving its position as the best informed agency will the Authority be able to tell where the line between the intrinsically dangerous and the nondangerous should be drawn. If it turns out at some time in the future, as a result of new discoveries, that other materials or other processes lend themselves to dangerous atomic developments, it is important that the Authority should be the first to know. At that time measures would have to be taken to extend the boundaries of safeguards.

5. Strategic Distribution of Activities and Materials—The activities entrusted exclusively to the Authority because they are intrinsically dangerous to security should be distributed throughout the world. Similarly, stockpiles of raw materials and fissionable materials should not be centralized.

6. Non-Dangerous Activities—Atomic research (except in explosives), the use of research reactors, the production of radioactive tracers by means of non-dangerous reactors, the use of such tracers, and to some extent the production of power should be open to nations and their citizens under reasonable licensing arrangements from the Authority. Denatured materials necessary for these activities should be furnished, under lease or other suitable arrangement by the Atomic Development Authority.

It should be an essential function of the Atomic Development Authority to promote to the fullest possible extent the peace-time benefits that can be obtained from the use of atomic energy.

It is necessary at all times to take advantage of the opportunity for promotion of decentralized and diversified national and private developments and of avoiding unnecessary concentration of functions in the Authority. It should, therefore, be a primary function to the Authority to encourage development by nations and private enterprise in the broad field of non-dangerous activities.

7. Definition of Dangerous and Non-Dangerous Activities—Although a reasonable dividing line can be drawn between the dangerous and the nondangerous, it is not hard and fast. Machinery should, therefore, be provided to assure constant examination and re-examination of the question, and to permit revision of the dividing line as changing conditions and new discoveries may require.

7 (a) Any plant dealing with uranium or thorium after it once reaches the potential of dangerous use must be not only subject to the most rigor-

ous and competent inspection by the international Authority, but its actual operation shall be under the management, supervision and control of the international Authority.

8. Inspection Activities—By assigning intrinsically dangerous activities exclusively to management by the Atomic Development Authority, the difficulties of inspection are thereby reduced to manageable proportions. For if the Atomic Development Authority is the only agency which may lawfully conduct the dangerous activities in the field of raw materials, primary production plants, and research in explosives, then visible operation by others than the Authority will constitute a danger signal.

The plan does not contemplate any systematic or large-scale inspection procedures covering the whole of industry. The delegation of authority for making inspections will have to be carefully drawn so that the inspection may be adequate for the needs and responsibilities of the Authority and yet not go beyond this point. Many of the inspection activities of the Authority should grow out of and be incidental to its other functions. An important measure of inspection will be those associated with the tight control of raw materials, for this is one of the keystones of the plan. The continuing activities of prospecting, survey and research in relation to raw materials will be designed not only to serve the affirmative development functions of the agency but also to assure that no surreptitious operations are conducted in the raw materials field by nations or their citizens. Inspection will also occur in connection with the licensing functions of the Authority. Finally, a means should be provided to enable the international organization to make special "spot" investigations of any suspicious national or private activities.

9. Personnel—The personnel of the Atomic Development Authority should be recruited on a basis of proven competence but also so far as possible on an international basis, giving much weight to geographical and national distribution. Although the problem of recruitment of the high-quality personnel required for the top executive and technical positions will be difficult, it will certainly be far less difficult than the recruitment of the similarly high-quality personnel that would be necessary for any purely policing organization.

10. Negotiation Stage—The first step in the creating of the system of control is the spelling out in comprehensive terms of the functions, responsibilities, authority, and limitations of the Atomic Development Authority. Once a Charter for the agency has been written, and adopted, the Authority and the system of control for which it will be responsible will require time to become fully organized and effective. The plan of control will therefore have to come into effect in successive stages. These should be specifically fixed in the Charter or means should be otherwise set forth in

the Charter for transitions from one stage to another, as contemplated in the resolution of the U. N. Assembly which created this Commission.

11. Disclosures—In the deliberation of the United Nations Commission on Atomic Energy, the United States must be prepared to make available the information essential to a reasonable understanding of the proposals which it advocates. Further disclosures must be dependent, in the interests of all, upon the effective ratification of this treaty. If and when the Authority is actually created, the United States must then also be prepared to make available other information essential to that organization or the performance of its functions. And as the successive stages of international control are reached, the United States must further be prepared to yield, to the extent required by each stage, national control of activities in this field to the international agency.

12. International Control—There will be questions about the extent of control allowed to national bodies, should an international body be established and in this respect, it is believed that any control by an atomic energy authority set up by any state should to the extent necessary for the effective operation of the international control system be subordinate to direction and absolute dominion on the part of the international authority. It will readily be seen at this time that this is not an endorsement or disapproval of the creation of national authorities, or a definition of their jurisdiction. This problem will be before the Commission and it should deal with a clear separation of duties and responsibilities of such state authorities, with the purpose of preventing possible conflicts of jurisdiction.

Harry S. Truman

---

23. "What Is the Atom's Industrial Future?", March 8, 1947

Man has always been a parasite of the sun. Our bodies run on solar power picked up by plants. Our industries sometimes use the mechanical device of falling water to tap sunpower. More often they run on the stored solar energy of coal and oil—stored by prehistoric plants which used radiation from the sun to pack chemical elements into unstable arrangements. The energy is released when they rearrange into more stable forms.

* A Different Source—On December 2, 1942, we tapped an entirely

This text appeared in *Business Week*, March 8, 1947, pp. 21–22.

new source of energy, independent of the sun. This was the energy stored, before the suns were formed, when subatomic particles were packed together to form two unstable heavy elements, uranium and thorium. On that day, physicists at the University of Chicago demonstrated that several tons of uranium, embedded with the proper spacing in several tons of graphite, would generate heat and continue to do so in a controllable, self-sustained process.

The applications of that discovery have already revolutionized one major industry—war—and have fundamentally changed the balance of international strength. Outside of those fields, it's true, most people are still inclined to act as though atomic energy were a private problem of statesmen, soldiers, and physicists.

* Effect on Industry—But in recent months, alert industrialists have begun to realize that such a basic change, in an area that runs as close to the heart of every industry and technology as does fuel and power supply, is sure to have far-reaching effects. This is even more certain in view of the fact that exploitation of the new development has the full resources of the government behind it—to the tune, currently, of some half-billion dollars a year.

Today, no prudent businessman, no prudent engineer dares make plans or decisions reaching more than about five years into the future without at least weighing the possibility that the basis of his planning may be altered or upset by the commercialization of discoveries about the atom.

* Problems Posed—One immediate effect of this realization is to add a lot of question marks to problems such as plant location, process design, raw material selection. From the technical side, firm answers to most questions on the economics of the atom are several years away. What information exists is veiled in secrecy. And political factors, national or international, may delay or modify the commercialization of nuclear processes.

But that there will be large economic effects, no one questions.

Among technical and businessmen close to the atomic work, there's increasing agreement on the early applications to be expected. These conclusions are frankly speculative. Competent men can be found who will insist that the expected applications are overrated and the time factors underestimated. But business decisions which ignore these predictions should be taken in the clear realization that a good deal of competent opinion is being overridden.

I. Power Generation

Commercial production of electric power from atomic engines is only about five years away. That's the informed assumption today. Experimental production of power on a substantial scale will be under way within a couple of years in at least three places—Oak Ridge, Schenectady, and Chicago.

Five years is much sooner than was being predicted a few months ago. The figure reflects promising results of preliminary investigations connected with the design of the three power piles on the drafting boards.

* Question of Temperature—energy production from atomic piles, such as those at the Hanford (Wash.) plutonium plant, has been a fact for several years. But it's released by the pile in the form of heat. And so far, it has been produced at too low a temperature to be used for anything except to heat the Columbia River by several degrees.

The new power piles will be designed to operate at temperatures similar to those of a modern, high-pressure steam plant. Some fluid such as molten bismuth, sodium-potassium alloys, or others which can reach high temperatures without creating high pressures, will be heated in the pile; the hot metal will generate high-pressure steam in a heat exchanger to run more or less conventional generating equipment.

New piles at Clinton Laboratory in Oak Ridge and the Knolls Laboratory in Schenectady are specifically designed for power production. The Argonne laboratory's new pile at Chicago, though intended for general experimentation, will run at high temperature and will incorporate a heat exchanger. And it's a safe bet that any of the future piles being planned—in the U.S., in Canada, in England—will be designed to utilize the energy produced.

* Competitive Position—Cost will not be an obstacle to the use of atom-fueled electricity. Only official word so far on cost is an estimate of 8 mills per kwh. It was submitted by R. C. Tolman, scientific adviser of the U.S. delegate to the international committee on control of atomic energy. This figure is some 30% higher than the cost of coal-generated power in areas where coal is plentiful. But it would make atomic power economical in any area where coal costs more than about $10 a ton.

If, as now appears likely, Europe and England face permanent shortages of coal, then atomic power, even at 8 mills, could be a matter of economic life or death wherever no large blocks of hydro power are available.

A great many engineers and physicists in atom work are convinced, however, that this official figure is far too high, was deliberately made

ultraconservative. They think they're working on a power source which, right from the start, will be competitive on a cost basis with any other fuel.
* Forecast—There are highly respected scientists who predict privately that within 20 years substantially all central station power will be drawn from atomic sources.

If this prediction is even partly right the effects on the coal and oil industries, the railroads, the industries using oil and coal as raw materials, and the industries tributary to all these groups are incalculable.
* Complications—A complicating factor in determining actual power cost is the question of byproducts. An operating pile always produces valuable radioactive materials. These could be an additional source of income (of course, the cost of the labor necessary to handle the fission products would have to be deducted). Another income source might be the use of the pile for experimental, metallurgical, or chemical purposes.

Thus, pricing of atomic power is likely to involve some of the complications encountered in distributing the costs of a multipurpose hydroelectric project, such as TVA.
* No Lack of Ore—Supply of atomic fuel does not appear to be any obstacle either. Uranium and thorium, the two known fuels, are rather plentiful. Known world resources of good ore are said to have sufficient power potential to meet the world's total power needs all by themselves for something like 20 years. Any exploration for uranium has been intensive for only a few years. Extraction from low-grade sources—granite, shales, seawater—has hardly been studied at all.

Statistically, uranium is a fairly common element, scarcer than copper but more plentiful than tungsten, zinc, or lead. Thorium is about three times as plentiful as uranium.

Once the technical problems of building commercial power plants are worked out, installation where they're needed will probably be able to go forward rather rapidly if the international situation permits. The fissionable material now being manufactured and stockpiled for military purposes can, if it is released, provide the initial material for power piles.

There are some other characteristics, peculiar to atomic power, which seem at the moment highly significant in assessing the industrial effects of commercialized atomic energy.
* Limitations—Atomic energy will prove most suitable for central-station power installations. The necessity for shielding reactors and for disposing of fission products seems to rule out most uses of atomic engines in vehicles. Exceptions might include large naval vessels, where an unlimited cruising radius is desirable; submarines, where the absence of air consumption would make indefinite submersion possible; and perhaps very long range or interplanetary rockets.

The central stations can be of two types: primary piles, similar to those at Hanford, and secondary reactors. Primary piles produce both heat (convertible into power) and secondary or "manufactured" fissionable materials, such as plutonium. Engineers talk in terms of 500,000 kw. or more for such primary reactors. Secondary reactors are units which use the secondary fissionable materials—probably in denatured form—as the source of energy. Potential output of such a reactor is perhaps 20,000 kw. to 100,000 kw. (BW-Oct. 25 '46, p24).

* Variable Output—Technically, the power output of any atomic device is extremely flexible, can be varied easily from a few watts to the level of an atomic bomb. Maximum output of an atomic power plant is limited only by the capacity of its heat-transfer system.

But from an economic standpoint, a high load factor is imperative, because capital investment is by far the largest element of cost. Once a plant was fueled up, cost of fuel replenishment would be a minor factor.

* Flexibility—Most novel feature of an atomic power plant is that it is independent of transportation. Because its fuel input is measured in pounds rather than tons, it can be located wherever power is needed, no matter how inaccessible the spot.

Thus, it can have a profound effect on the geography of all power-consuming heavy industries. It might become economic, for instance, to refine aluminum near the bauxite deposits, rather than to transport bauxite to a good power supply, as is done today.

II. Transmutation

It's as a power source that nuclear fission can be expected to have its immediate and massive economic effects. But it also opens the way to do things to matter, to change its character, in ways that were never possible before—or only on a minute laboratory scale. In the atomic pile, energies never before available can be turned loose on the structure of matter.

* Alchemist's Dream—One thing already obvious is that the old dream of transmutation of the elements is now an accomplished fact, and on a significant scale. Transmutation has been possible for years in the laboratory, using cyclotrons and other types of "guns" that shoot subatomic particles. But always before it's been on a scale too small to have industrial significance.

The biggest job of transmutation to date is the manufacture of fissionable material itself. One such material occurs in nature, the scarce 235 form of uranium. By using the energies of fission of uranium 235, the common 238 form of uranium is transmuted into the synthetic fissionable element, plutonium. Similarly, thorium, not readily fissionable, can be transmuted into a fissionable form of uranium (U235).

It's quite possible that in the future some of the now rare but important elements may be most easily obtained, not by prospecting for them, but by manufacturing them out of some other element.

* New Tools—Even more significant is the manufacture of radioactive isotopes of common elements. Most elements never occur in radioactive form in nature. But radioactive forms of nearly every element can be made in quantity in an atomic pile. They have exactly the same chemical properties as the natural elements. But they possess the additional property of emitting radioactive waves, just as radium does.

The research and medical value of radio-isotopes has been well publicized. But their potential usefulness in many industrial processes is not as widely realized. Two types of use are promising—as radiation sources and as tracers.

* Radiation Uses—As a source of penetrating, high-energy radiation, a package of radioactive material provides a comparatively inexpensive way of doing many inspection and other jobs that now call for bulky x-ray apparatus.

A cheap and convenient source of radiation may also have revolutionary uses in the chemical industries. In many chemical reactions, radiation has a catalytic effect, permitting reactions at low temperatures which otherwise would occur only at high temperatures or not at all. This effect has been little studied, except in photography. But several oil companies are now investigating its application to the cracking of petroleum.

Even such prosaic uses as sterilization of sewage and ionization of the air around a moving belt to dissipate static charges are being seriously considered.

* Tracer Uses—Radioactive isotopes get their value as tracers from the fact that even microscopic quantities can easily be detected by their radiation. That is, the radioactive waves which they emit can be easily recorded on instruments.

The ability to identify a particular bit of material has great research value in such a problem, say, as studying the exchange of metal between the two surfaces of a bearing.

A bit of radioactive copper included in the bronze of some inaccessible bearing would make it possible to detect wear in the bearing by a radiation check of the lubricating oil.

Cavitation (vacuum formation) around turbine blades might be detected in the same way.

Proportions of alloying materials could be checked without the need for a chemical analysis.

The possibilities are endless.

Production of radio-isotopes is still so small as to limit their use to laboratories. But they'll be available to industry within a year or two. Any wide use of power piles would make radio-isotopes a routine industrial tool.

* Effect on Crystal Structure—There is still another possible application of atomic energy to the field of metallurgy. The crystalline structure of materials has been found to change markedly when they are subjected to bombardment from neutrons in an atomic pile.

Since the physical properties of most metals depend on their crystal structure, many investigators think that neutron bombardment could be used as a new form of "heat treatment," producing properties not obtainable by the conventional annealing and tempering.

III. The Future

As now conceived, atomic energy merely substitutes an atomic pile for a conventional steam plant in developing the heat required to generate power. But atomic energy itself is an electrical force. Some day a means may be found to harness this force directly, instead of using it to generate heat which must be reconverted into power. Today this thought is little more than a dream in researchers' eyes—but so was atomic fission a decade ago.

In the offing is the possibility that the tool of uranium fission may lead to other energy-releasing nuclear changes. In theory, there are many such changes. The stable elements are those of middle weight in the neighborhood of iron; when heavy elements break down toward this middle level or when light elements combine to form middle-weight ones, energy should be released.

* Hydrogen Research—It is known that promising research is already under way on energy from hydrogen. When atoms of hydrogen combine to form helium, much more energy is freed than by the fission of the same quantity of uranium. The energy of the sun, itself, is believed to come from this combination.

Standard theory is that the hydrogen-to-helium transmutation can occur only at temperatures such as those in the sun—millions of degrees. If this is correct, it would seem that a chain reaction, generating heat at this temperature faster than it could be dissipated, could only be maintained in a body somewhere near the size of the sun.

* Superbomb?—Thus hydrogen hardly looks promising as a controlled power source. But the announcement of research on hydrogen is widely

taken to mean that a uranium bomb could be used to produce a flash of sun-scale temperature that would set off an atomic reaction in a quantity of hydrogen—perhaps resulting in a bomb a hundred or a thousand times as powerful as the outmoded plutonium affair of World War II.

---

24. President Dwight D. Eisenhower's "Atoms for Peace" Address to the United Nations General Assembly, December 8, 1953

Madame[1] President, Members of the General Assembly:

When Secretary General Hammarskjold's invitation to address this General Assembly reached me in Bermuda, I was just beginning a series of conferences with the Prime Ministers and Foreign Ministers of Great Britain and of France. Our subject was some of the problems that beset the world.

During the remainder of the Bermuda Conference, I had constantly in mind that ahead of me lay a great honor. That honor is mine today as I stand here, privileged to address the General Assembly of the United Nations.

At the same time that I appreciate the distinction of addressing you, I have a sense of exhilaration as I look upon this assembly.

Never before in history has so much hope for so many people been gathered together in a single organization. Your deliberations and decisions during these somber years have already realized part of those hopes.

But the great tests and the great accomplishments still lie ahead. And in the confident expectation of those accomplishments, I would use the office which, for the time being, I hold, to assure you that the Government of the United States will remain steadfast in its support of this body. This we shall do in the conviction that you will provide a great share of the wisdom, of the courage and the faith which can bring to this world lasting peace for all nations and happiness and well-being for all men.

Clearly, it would not be fitting for me to take this occasion to present to you a unilateral American report on Bermuda. Nevertheless, I assure you that in our deliberations on that lovely island we sought to invoke those

This text appeared in the *Congressional Record*, vol. 100, January 7, 1954, pp. 61–63.
1. The President's opening words referred to Mme. Vijaya Pandit, president of the United Nations General Assembly.

same great concepts of universal peace and human dignity which are so cleanly etched in your Chapter.

Neither would it be a measure of this great opportunity merely to recite, however hopefully, pious platitudes.

I therefore decided that this occasion warranted my saying to you some of the things that have been on the minds and hearts of my legislative and executive associates and on mine for a great many months—thoughts I had originally planned to say primarily to the American people.

I know that the American people share my deep belief that if a danger exists in the world, it is a danger shared by all—and equally, that if hope exists in the mind of one nation, that hope should be shared by all.

Finally, if there is to be advanced any proposal designed to ease, even by the smallest measure, the tensions of today's world, what more appropriate audience could there be than the members of the General Assembly of the United Nations?

Language of Atomic Warfare

I feel impelled to speak today in a language that, in a sense, is new—one, which I, who spent so much of my life in the military profession, would have preferred never to use.

That new language is the language of atomic warfare.

The atomic age has moved forward at such a pace that every citizen of the world should have some comprehension, at least in comparative terms, of the extent of this development, of the utmost significance to every one of us. Clearly, if the people of the world are to conduct an intelligent search for peace, they must be armed with the significant facts of today's existence.

My recital of atomic danger and power is necessarily stated in United States terms, for these are the only incontrovertible facts that I know. I need hardly point out to this assembly, however, that this subject is global, not merely national in character.

On July 16, 1945, the United States set off the world's first atomic test explosion. Since that date in 1945, the United States of America has conducted forty-two test explosions.

Atomic bombs today are more than twenty-five times as powerful as the weapons with which the atomic age dawned, while hydrogen weapons are in the ranges of millions of tons of TNT equivalent.

Today, the United States' stockpile of atomic weapons, which, of course, increases daily, exceeds by many times the explosive equivalent of the total of all bombs and all shells that came from every plane and every gun in every theatre of war through all the years of World War II.

A single air group, whether afloat or land based, can now deliver to any reachable target a destructive cargo exceeding in power all the bombs that fell on Britain in all of World War II.

In size and variety the development of atomic weapons has been no less remarkable. This development has been such that atomic weapons have virtually achieved conventional status within our armed services. In the United States services, the Army, the Navy, the Air Force and the Marine Corps are all capable of putting this weapon to military use.

But the dread secret and the fearful engines of atomic might are not ours alone.

In the first place, the secret is possessed by our friends and Allies, Great Britain and Canada, whose scientific genius made a tremendous contribution to our original discoveries and the designs of atomic bombs.

The secret is also known by the Soviet Union.

The Soviet Union has informed us that, over recent years, it has devoted extensive resources to atomic weapons. During this period, the Soviet Union has exploded a series of atomic devices, including at least one involving thermo-nuclear reactions.

Monopoly Is Now Ended

If at one time the United States possessed what might have been called a monopoly of atomic power, that monopoly ceased to exist several years ago. Therefore, although our earlier start has permitted us to accumulate what is today a great quantitative advantage, the atomic realities of today comprehend two facts of even greater significance.

First, the knowledge now possessed by several nations will eventually be shared by others, possibly all others.

Second, even a vast superiority in numbers of weapons, and a consequent capability of devastating retaliation, is no preventive, of itself, against the fearful material damage and toll of human lives that would be inflicted by surprise aggression.

The free world, at least dimly aware of these facts, has naturally embarked on a large program of warning and defense systems. That program will be accelerated and expanded.

But let no one think that the expenditure of vast sums for weapons and systems of defense can guarantee absolute safety for the cities and the citizens of any nation. The awful arithmetic of the atomic bomb does not permit of such an easy solution. Even against the most powerful defense, an aggressor in possession of the effective minimum number of atomic bombs or a surprise attack could probably place a sufficient number of his bombs on the chosen targets to cause hideous damage.

Should such an atomic attack be launched against the United States, our reaction would be swift and resolute. But for me to say that the defense capabilities of the United States are such that they could inflict terrible losses upon an aggressor—for me to say that the retaliation capabilities of the United States are so great that such an aggressor's land would be laid waste—all this, while fact, is not the true expression of the purpose and the hope of the United States.

To pause there would be to confirm the hopeless finality of a belief that two atomic colossi are doomed malevolently to eye each other indefinitely across a trembling world. To stop there would be to accept helplessly the probability of civilization destroyed—the annihilation of the irreplaceable heritage of mankind handed down to us generation from generation—and the condemnation of mankind to begin all over again the age-old struggle upward from savagery toward decency and right and justice.

No Victory in Desolation

Surely no sane member of the human race could discover victory in such desolation. Could anyone wish his name to be coupled by history with such human degradation and destruction?

Occasional pages of history do record the faces of the "Great Destroyers" but the whole book of history reveals mankind's never-ending quest for peace and mankind's God-given capacity to build.

It is with the book of history, and not with isolated pages, that the United States will ever wish to be identified. My country wants to be constructive, not destructive. It wants agreements, not wars, among nations. It wants, itself, to live in freedom and in the confidence that the people of every other nation enjoy equally the right of choosing their own way of life.

So my country's purpose is to help us move out of the dark chamber of horrors into the light, to find a way by which the minds of men, the hopes of men, the souls of men everywhere, can move forward toward peace and happiness and well-being.

In this quest, I know that we must not lack patience.

I know that in a world divided, such as ours today, salvation cannot be attained by one dramatic act.

I know that many steps will have to be taken over many months before the world can look at itself one day and truly realize that a new climate of mutually peaceful confidence is abroad in the world.

But I know, above all else, that we must start to take these steps—now.

The United States and its Allies, Great Britain and France, have, over the past months, tried to take some of these steps. Let no one say that we shun the conference table.

On the record has long stood the request of the United States, Great Britain and France, to negotiate with the Soviet Union the problems of a divided Germany.

On that record has long stood the request of the same three nations to negotiate an Austrian peace treaty.

On the same record still stands the request of the United Nations to negotiate the problems of Korea.

Conference with the Russians

Most recently, we have received from the Soviet Union what is in effect an expression of willingness to hold a four-power meeting. Along with our Allies, Great Britain and France, we were pleased to see that this note did not contain the unacceptable preconditions previously put forward.

As you already know from our joint Bermuda communique, the United States, Great Britain and France have agreed promptly to meet with the Soviet Union.

The Government of the United States approaches this conference with hopeful sincerity. We will bend every effort of our minds to the single purpose of emerging from that conference with tangible results toward peace—the only true way of lessening international tension.

We never have, we never will, propose to suggest that the Soviet Union surrender what is rightfully theirs.

We will never say that the peoples of Russia are an enemy with whom we have no desire ever to deal or mingle in friendly and fruitful relationship.

On the contrary, we hope that this coming conference may initiate a relationship with the Soviet Union which will eventually bring about a free intermingling of the peoples of the East and of the West—the one sure, human way of developing the understanding required for confident and peaceful relations.

Instead of the discontent which is now settling upon Eastern Germany, occupied Austria and the countries of Eastern Europe, we seek a harmonious family of free European nations, with none a threat to the other, and least of all a threat to the peoples of Russia.

Beyond the turmoil and strife and misery of Asia, we seek peaceful opportunity for these peoples to develop their natural resources and to elevate their lot.

These are not idle words of shallow vision. Behind them lies a store of nations lately come to independence, not as a result of war but through free grant or peaceful negotiation. There is a record already written of assistance gladly given by nations of the West to needy peoples and to those suffering the temporary effects of famine, drought and natural disaster.

These are deeds of peace. They speak more loudly than promises or protestations of peaceful intent.

Would Explore Every Channel

But I do not wish to rest either upon the reiteration of past proposals or the restatement of past deeds. The gravity of the time is such that every new avenue of peace, no matter how dimly discernible, should be explored.

There is at least one new avenue of peace which has not yet been well explored—an avenue now laid out by the General Assembly of the United Nations.

In its resolution of November 18, 1953, this General Assembly suggested—and I quote—"that the Disarmament Commission study the desirability of establishing a subcommittee consisting of representatives of the powers principally involved, which should seek, in private, an acceptable solution—and report such a solution to the General Assembly and to the Security Council not later than 1 September, 1954."

The United States, heeding the suggestion of the General Assembly of the United Nations, is instantly prepared to meet privately with such other countries as may be "principally involved," to seek "an acceptable solution" to the atomic armaments race which overshadows not only the peace but the very life of the world.

We shall carry into these private or diplomatic talks a new conception.

The United States would seek more than the mere reduction or elimination of atomic materials for military purposes.

It is not enough to take this weapon out of the hands of the soldiers. It must be put into the hands of those who will know how to strip its military casing and adapt it to the arts of peace.

The United States knows that if the fearful trend of atomic military build-up can be reversed, this greatest of destructive forces can be developed into a great boon for the benefit of all mankind.

The United States knows that peaceful power from atomic energy is no dream of the future. That capability, already proved, is here now—today. Who can doubt, if the entire body of the world's scientists and engineers had adequate amounts of fissionable material with which to test and develop their ideas, that this capability would rapidly be transformed into universal, efficient and economic usage.

To hasten the day when fear of the atom will begin to disappear from the minds of people and the governments of the East and West there are certain steps that can be taken now.

I therefore make the following proposals:

The governments principally involved to the extent permitted by elementary prudence, to begin now and continue to make joint contributions from their stockpiles of normal uranium and fissionable materials to an international atomic energy agency. We would expect that such an agency would be set up under the aegis of the United Nations.

The ratios of contributions, the procedures and other details would properly be within the scope of the "private conversations" I have referred to earlier.

The United States is prepared to undertake these explorations in good faith. Any partner of the United States acting in the same good faith will find the United States a not unreasonable or ungenerous associate.

Undoubtedly initial and early contributions to this plan would be small in quantity. However, the proposal has the great virtue that it can be undertaken without irritations and mutual suspicions incident to any attempt to set up a completely acceptable system of world-wide inspection and control.

The atomic energy agency could be made responsible for the impounding, storage and protection of the contributed fissionable and other material. The ingenuity of our scientists will provide special, safe conditions under which such a bank of fissionable material can be made essentially immune to surprise seizure.

The more important responsibility of this atomic energy agency would be to devise methods whereby this fissionable material would be allocated to serve the peaceful pursuits of mankind. Experts would be mobilized to apply atomic energy to the needs of agriculture, medicine and other peaceful activities. A special purpose would be to provide abundant electrical energy in the power-starved areas of the world. Thus the contributing powers would be dedicating some of their strength to serve the needs rather than the fears of mankind.

Outlines Plan for Congress

The United States would be more than willing—it would be proud—to take up with others "principally involved" the development of plans whereby such peaceful use of atomic energy would be expedited.

Of those "principally involved" the Soviet Union must, of course, be one.

I would be prepared to submit to the Congress of the United States, and with every expectation of approval, any such plan that would:

First, encourage world-wide investigation into the most effective peacetime uses of fissionable material;

Second, begin to diminish the potential destructive power of the world's atomic stockpiles;

Third, allow all peoples of all nations to see that, in this enlightened age, the great powers of the earth, both of the East and of the West, are interested in human aspirations first rather than building up the armaments of war;

Fourth, open up a new channel for peaceful discussion and initiate at least a new approach to the many difficult problems that must be solved in both private and public conversations if the world is to shake off the inertia imposed by fear and is to make positive progress toward peace.

Against the dark background of the atomic bomb, the United States does not wish merely to present strength, but also the desire and the hope for peace.

The coming months will be fraught and fateful decisions. In this Assembly, in the capitals and military headquarters of the world; in the hearts of men everywhere, be they governed or governors, may they be the decisions which will lead this world out of fear and into peace.

To the making of these fateful decisions, the United States pledges before you—and therefore before the world—its determination to help solve the fearful atomic dilemma—to devote its entire heart and mind to find the way by which the miraculous inventiveness of man shall not be dedicated to his death, but consecrated to his life.

I again thank the delegates for the great honor they have done me in inviting me to appear before them and in listening to me so courteously. Thank you.

25. Excerpts from President Eisenhower's Press Conference, March 31, 1954

The following questions by news correspondents and answers by Lewis L. Strauss, Chairman, AEC, came at the conclusion of his prepared statement, which is attached.

Q. Mr. Chairman, you said that this particular explosion was not out of control. But is it possible that in any series of tests that a hydrogen explosion or series of them could get out of control?

A. I am informed by the scientists that that is impossible.

Q. Admiral Strauss, yesterday, at his news conference, Secretary of Defense Wilson said the results of the March 1st test—is the one he was referring to—was unbelievable. Would you care to comment on that?

A. No, I don't think I should comment on that. The use of that adjective, I think, was played up beyond the point where the Secretary intended it. I don't know what is meant by "unbelievable" and I would rather not comment.

Q. Mr. Chairman, do you intend to imply by the last paragraph in this statement that the work on the weapon phase of the atom is reaching a completion; that we are approaching a point where pursuit of this will no longer yield very large profits, and that we will, therefore, turn our research power to the peaceful application?

A. I think the answer to that is this: The Military have certain requirements. The Commission is engaged in attempting to fill those requirements. The ability of the Commission to devote attention and fissionable material to peaceful requirements, peaceful needs, is always junior to the defense needs, by definition of the Act itself. The result of these tests has brought us very much nearer to the day of the satisfaction of military requirements, put us within sight of them, so that we can see the ability to proceed aggressively with the peacetime development of power to an extent that we were not able to before the tests.

Q. Thank you, sir.

Q. Admiral Strauss, can you go beyond this statement and describe the area of the blast, the effectiveness of the blast, and give a general description of what actually happened when the H-bomb went off?

This text appeared in Eisenhower Press Conference, March 31, 1954; *New York Times*, April 1, 1954.

A. The area of the blast, would be about—

THE PRESIDENT: Why not depend on these pictures they are all going to see?

A. (Mr. Strauss) I understand you are going to see a film, a picture, of the 1952 shot. The area, if I were to describe it specifically, would be translatable into the number of megatons involved, which is a matter of military secrecy.

The effects, you said the effectiveness—I don't know exactly what you meant by that, sir, so I don't know how to answer it.

Q. Well, I don't mean in the percentage of the effectiveness of or the efficiency of the blast itself. But many people in Congress, I think many elsewhere, have been reaching out and grasping for some information as to what happens when the H-bomb goes off, how big is the area of destruction in its various stages; and what I am asking you for now is some enlightenment on that subject.

A. Well, the nature of an H-bomb, Mr. Wilson, is that, in effect, it can be made to be as large as you wish, as large as the military requirement demands, that is to say, an H-bomb can be made as—large enough to take out a city.

(A chorus of "What?")

To take out a city, to destroy a city.

Q. How big a city?

A. Any city.

Q. Any city, New York?

A. The metropolitan area, yes.

(With reference to the foregoing, Mr. Strauss added later that he meant "put out of commission," not "to destroy.")

Q. Mr. Chairman, may I ask this specific question: If you were to make a comparison, duplicating the explosion that occurred at Eniwetok, with this building in which we are right now as the center, what would be left of this city of Washington?

A. Well I couldn't say, because the precise measurements of these two shots have not been completely calibrated. It may be as many as a month or two before I know the answer to it. It would be a very extensive—

Q. Will you provide that answer at some time, sir?

A. I won't make a definite commitment, but I would certainly like to.

IV

The Hydrogen Bomb

*I*n August 1949 the first successful Soviet test of an atomic bomb, nick-named "Joe I" after Stalin, ended the American monopoly on nuclear weapons and inaugurated the era of proliferation. The American response was to launch a crash program to develop an even more fearful thermonuclear, or hydrogen, weapon based upon the fusing together at extremely high temperatures of light elements, as opposed to the fission of heavy ones. Although the idea of a fusion weapon had been broached by the Hungarian physicist Edward Teller in 1942, relatively little work had been done.

Despite opposition by Oppenheimer and the General Advisory Committee of the Atomic Energy Commission, Truman decided to approve the development of the new weapon upon the advice of his military advisers. The achievement of a successful hydrogen bomb did not occur until 1954, due less to the opposition of Oppenheimer than to technical and design errors of Teller and the complexity of the fusion process, which required coupling by x-rays of the primary and secondary parts of the weapon. Whereas the early atomic bombs had a yield of approximately twenty kilotons of TNT equivalent, the hydrogen bomb yield immediately reached the fifteen megaton range; the largest weapon was exploded in 1961 by the Soviet Union, with a yield of nearly sixty megatons.

On the development of the hydrogen bomb, see Herbert York, The Advisors: Oppenheimer, Teller and the Superbomb *(San Francisco: W. H. Freeman, 1976),* which includes the previously classified General Advisory Committee report. More recently the 1954 history of the hydrogen bomb project by Hans Bethe has been published in the Los Alamos Science *(Fall 1982)* as *"Comments on the History of the H-Bomb."* In 1979 a journalist named Howard Morland attempted to publish a description of the hydrogen *bomb "secret" in the* Progressive *magazine, and the U.S. government took him to court and delayed publication. For the details of the case, see Alexander De Volpi et al.,* Born Secret: The H-Bomb, the Progressive Case, and National Security *(Oxford: Pergamon Press, 1981).*

---

26. "Political Implications of Detonation of Atomic Bomb by the U.S.S.R.," Department of State, Policy Planning Staff PPS/58, August 16, 1949

Top Secret

The Problem

1. To determine the political implications if this Government could know with certainty when the U.S.S.R. detonates an atomic bomb.

Analysis and Conclusions

The Department of State obviously cannot pass on the question whether scientific techniques or equipment can be developed to detect the explosion by the U.S.S.R. of an atomic bomb, and it cannot express judgment as between competing demands for research and development funds. It is clear, however, that "only if a high degree of certainty can be placed on systems of detection, would this Government be warranted in basing policy decisions on intelligence derived from them."

Definite knowledge by this Government of the explosion by the U.S.S.R. of its first bomb is considered by the Department to be important for the following reasons:

(1) It would have a steadying effect on the American people and give them a sense of security if this Government could give assurance that the U.S.S.R. probably could not, without our knowledge, have a bomb or bombs for any length of time. With this knowledge, the Government would be able to combat intelligently defeatist or irrational attitudes arising from uncertainty as to whether the U.S.S.R. was capable of using atomic bombs, and would be in a position to refute with conviction false claims or rumors.

This text is found in Foreign Relations of the United States, *1949*, vol. 1 (Washington, D.C., 1976) pp. 514–15.

(2) It would be of the utmost importance for us to know when the U.S.S.R. has successfully tested a bomb in order to anticipate and counter possible changes in Soviet foreign policy which might result therefrom, and to know whether a shift in its foreign policy was the result of the possession of atomic bombs. We cannot know whether the U.S.S.R. would make the knowledge public if it did possess the atomic bomb; however, we would be in a position to know the truth of what the U.S.S.R. said publicly.

(3) The Soviet possession of a bomb or bombs may require a reevaluation of U.S. policy in the United Nations in our efforts to obtain effective international control.

(4) Most of the free nations of the world are inclined at present to cooperate with the United States in view of the threat of Soviet aggression. A belief that we are now the sole possessor of atomic bombs and that the U.S.S.R. has none probably tends to increase their desire to collaborate with us and also their sense of safety in doing so. This tendency would probably be reinforced even further by certain knowledge that the U.S.S.R. does not possess the bomb and that we would have means of knowing if it and when it did come into possession of the bomb. However, it is realized that knowledge that the U.S.S.R. did in fact possess the bomb also might tend to incline third countries toward a position of neutrality between the United States and the U.S.S.R.

(5) If at some later time we should learn with certainty that the U.S.S.R. did possess the atomic bomb, this knowledge would be of importance in reevaluating the necessity for precautionary measures to reduce U.S. vulnerability to atomic attack. However, this is a matter of primary concern to the NME. (National Military Establishment)

---

27. Statement by the President on Announcing the First Atomic Explosion in the USSR, September 23, 1949

I believe the American people, to the fullest extent consistent with national security, are entitled to be informed of all developments in the field of atomic energy. That is my reason for making public the following information.

This text appeared in *Public Papers of the Presidents of the United States: Harry S. Truman, 1949* (Washington, D.C.: Government Printing Office, 1964), p. 485.

We have evidence that within recent weeks an atomic explosion occurred in the U.S.S.R.

Ever since atomic energy was first released by man, the eventual development of this new force by other nations was to be expected. This probability has always been taken into account by us.

Nearly 4 years ago I pointed out that "scientific opinion appears to be practically unanimous that the essential theoretical knowledge upon which the discovery is based is already widely known. There is also substantial agreement that foreign research can come abreast of our present theoretical knowledge in time." And, in the Three-Nation Declaration of the president of the United States and the prime ministers of the United Kingdom and of Canada, dated November 15, 1945, it was emphasized that no single nation could in fact have a monopoly of atomic weapons.

This recent development emphasizes once again, if indeed such emphasis were needed, the necessity for that truly effective enforceable international control of atomic energy which this Government and the large majority of the members of the United Nations support.

28. USAEC General Advisory Committee Minutes, October 28–30, 1949

MINUTES:

Seventeenth Meeting of the General Advisory Committee
to the U.S. Atomic Energy Commission,
October 28–30, 1949,
Washington, D.C.

FIRST SESSION
October 28, 1949

This Session was convened by the Chairman at 2:00 P.M. Those present were Dr. Oppenheimer, Dr. Buckley, Dr. DuBridge, Dr. Fermi, Dr. Rabi, Dr. Smith, the Secretary, and Mr. George F. Kennan, Counsellor of the Department of State. Dr. Seaborg was absent for all sessions of this meeting.

This text is found in United States Atomic Energy Commission General Advisory Committee Records, United States Department of Energy, Germantown, Maryland.

Mr. Kennan presented his views of the Russian situation and replied to a number of questions from Committee members. He left the meeting at 2:45 P.M.

The remainder of the Session, which Dr. Rowe joined, was devoted to continuing the discussion of our present policy and the role of atomic energy in present and future implementation of developing policy.

This Session was adjourned by 4:00 P.M. to permit the members to have informal discussions with Dr. H. A. Bethe of Cornell University and Dr. R. Serber of the Radiation Laboratory, Berkeley.

SECOND SESSION
October 29, 1949

This Session was convened at 9:30 A.M. Dr. Oppenheimer, Dr. Buckley, Dr. Conant, Dr. DuBridge, Dr. Fermi, Dr. Rabi, Dr. Rowe, Dr. Smith, the Secretary, and Mr. Tomei were present.

As the result of agreement that the proposal to initiate a program for the development of a super-bomb was the chief issue under the broad objective of examining whether the Commission is doing everything reasonably possible for it to do for the common defense and security, the discussion was directed toward a consideration of the details and possible consequences of such a program.

At 10:00 A.M., Commissioners Lilienthal, Dean, Smyth, and Strauss, the General Manager, the Deputy General Manager, the Director of Production, the Director of Reactor Development, The General Counsel, and the Executive Officer of the Program Council of the Commission joined the meeting, followed at 10:10 A.M. by the Director of Biology and Medicine and the Director of Military Application, and at 10:20 A.M. by the Director of Research. Mr. Tomei was absent for the remainder of this Session. Discussion of the super-bomb program continued with this group.

At 11:00 A.M., the Chairman of the Joint Chiefs of Staff, the Chairman and the Executive Secretary of the Military Liaison Committee, the Chairman of the Weapons Systems Evaluation Group, Lt. Gen. L. Norstad, and Rear Adm. W. S. Parsons joined the Committee to discuss the military factors relevant to the Commission's program. Commissioners Lilienthal, Dean, Smyth and Strauss, the General Manager, and the Director of Military Application of the Commission were also present. The purpose of this meeting was to explore the military implications of the Commission's present or future programs.

This Session was adjourned at 12:30 P.M.

THIRD SESSION

The Committee met with Dr. Malcolm C. Henderson, Deputy Director of Intelligence, AEC, at 2:10 P.M. to obtain background information on Russian activities as known to his office. Commissioners Lilienthal, Dean, Smyth, and Strauss were also present. Dr. Henderson left the meeting at 3:30 P.M. Discussion continued until adjournment of the Session at 5:00 P.M.

FOURTH SESSION

This Session was convened at 8:15 P.M. All Committee members (except Dr. Seaborg), the Secretary, and Mr. Tomei were present.

Further exploration of the super-bomb program was conducted. Before adjournment at 10:10 P.M. it was understood that drafts of specific positions would be the first order of business at the next Session.

FIFTH SESSION
October 30, 1949

This Session was convened at 9:05 A.M. Dr. Oppenheimer, Dr. Buckley, Dr. Conant, Dr. DuBridge, Dr. Fermi, Dr. Rabi, Dr. Rowe, Dr. Smith, the Secretary, and Mr. Tomei were present.

Preliminary positions on the super-bomb program were outlined, and other possibilities for extension of the Commission's program—reactors for production of freely available neutrons, increased fissionable material production, and the use of atomic weapons for tactical purposes—were discussed.

At 10:10 A.M., Commissioners Lilienthal, Dean, Pike, Smyth, and Strauss, the General Manager, and the General Counsel of the Commission joined the group for a summary by Dr. Oppenheimer of the views of the Committee at this time and for additional discussion. Mr. Tomei was absent until 12:15 P.M.

During this portion of this Session, the Committee agreed that the Commission should feel free to use as it wished the reports and written views it would receive from the Committee. It was the implication of this action that the members of the GAC would not discuss the Committee's views until freed to do so by action of the Commission. In addition, it was agreed that each member would avoid discussion of his personal opinion at least for a period of one week, the initial day of the proposed Los Alamos conference. This was done in order to assure the maximum freedom of action to the Commission in the matters under discussion.

The guests of the Committee left at 12:15 P.M.

After taking action to confirm the date of the next meeting as December 1, 2, and 3, 1949, the place as Washington, and the dates of the subsequent meeting as January 31 and February 1, 1950, the Committee constituted drafting groups for the preparation of papers. The papers thus prepared were edited and modified to consist of the following:

1. A report by the Chairman addressed to the Chairman of the AEC, consisting of a covering letter and two Parts, approved by all members of the Committee (except Dr. Seaborg, who was not present). The "majority" to which reference is made (Part II, page 4, last paragraph) included neither Dr. Fermi nor Dr. Rabi.
2. One attachment to the report approved and signed by Dr. Conant, Dr. Rowe, Dr. Smith, Dr. DuBridge, Dr. Buckley, and Dr. Oppenheimer.
3. A second attachment approved and signed by Dr. Fermi and Dr. Rabi.

It was agreed that in view of the nature of the documents they should bear a Top Secret classification.

The Minutes of the Sixteenth Meeting of the Committee were unanimously approved and this Meeting was adjourned at 3:00 P.M.

J. H. Manley
Secretary

29. USAEC General Advisory Committee Report on the "Super," October 30, 1949

GENERAL ADVISORY COMMITTEE
to the
U.S. ATOMIC ENERGY COMMISSION
Washington 25, D.C.

October 30, 1949

Dear Mr. Lilienthal:

At the request of the Commission, the seventeenth meeting of the General Advisory Committee was held in Washington on October 29 and

This text is found in United States Atomic Energy Commission General Advisory Committee Records, United States Department of Energy, Germantown, Maryland.

30, 1949 to consider some aspects of the question of whether the Commission was making all appropriate progress in assuring the common defense and security. Dr. Seaborg's absence in Europe prevented his attending this meeting. For purposes of background, the Committee met with the Counsellor of the State Department, with Dr. Henderson of AEC Intelligence, with the Chairman of the Joint Chiefs of Staff, the Chairman of the Military Liaison Committee, the Chairman of the Weapons Systems Evaluation Group, General Norstadt and Admiral Parsons. In addition, as you know, we have had intimate consultations with the Commission itself.

The report which follows falls into three parts. The first describes certain recommendations for action by the Commission directed toward the common defense and security. The second is an account of the nature of the super project and of the super as a weapon, together with certain comments on which the Committee is unanimously agreed. Attached to the report, but not a part of it, are recommendations with regard to action on the super project which reflect the opinions of Committee members.

The Committee plans to hold its eighteenth meeting in the city of Washington on December 1, 2, and 3, 1949. At that time we hope to return to many of the questions which we could not deal with at this meeting.

J. R. Oppenheimer
Chairman

UNITED STATES ATOMIC ENERGY COMMISSION
WASHINGTON, D.C. 20545
HISTORICAL DOCUMENT NUMBER 349

David E. Lilienthal
Chairman
U.S. Atomic Energy Commission
Washington 25, D.C.

PART I

(1) PRODUCTION. With regard to the present scale of production of fissionable material, the General Advisory Committee has a recommendation to make to the Commission. We are not satisfied that the present scale represents either the maximum or the optimum scale. We recognize the statutory and appropriate role of the National Military Establishment in helping to determine that. We believe, however, that before this issue can be settled, it will be desirable to have from the Commission a careful analysis of what the capacities are which are not now being employed. Thus we have in mind that an

acceleration of the program on beneficiation of low grade ores could well turn out to be possible. We have in mind that further plants, both separation and reactor, might be built, more rapidly to convert raw material into fissionable material. It would seem that some notion of the costs, yields and time scales for such undertakings would have to precede any realistic evaluation of what we should do. We recommend that the Commission undertake such studies at high priority. We further recommend that projects should not be dismissed because they are expensive but that their expense be estimated.

(2)   TACTICAL DELIVERY. The General Advisory Committee recommends to the Commission an intensification of efforts to make atomic weapons available for tactical purposes, and to give attention to the problem of integration of bomb and carrier design in this field.

(3)   NEUTRON PRODUCTION. The General Advisory Committee recommends to the Commission the prompt initiation of a project for the production of freely absorbable neutrons. With regard to the scale of this project the figure [deleted] per day may give a reasonable notion. Unless obstacles appear, we suggest that the expediting of design be assigned to the Argonne National Laboratory.

With regard to the purposes for which these neutrons may be required, we need to make more explicit statements. The principal purposes are the following:

(a) The production of U-233.
(b) The production of radiological warfare agents.
(c) Supplemental facilities for the test of reactor components.
(d) The conversion of U-235 to plutonium.
(e) A secondary facility for plutonium production.
(f) The production of tritium (1) for boosters, (2) for super bombs.

We view these varied objectives in a quite different light. We have a great interest in the U-233 program both for military and for civil purposes. We strongly favor, subject to favorable outcome of the 1951 Eniwetok tests, the booster program. With regard to radiological warfare, we would not wish to alter the position previously taken by our Committee. With regard to the conversion to plutonium, we would hardly believe that this alone could justify the construction of these reactors, though it may be important should unanticipated difficulties appear in the U-233 and booster programs. With regard to the use of tritium in the super bomb, it is our unanimous hope that this will not prove necessary. It is the opinion of the majority that the super program itself should not be undertaken and that the Commission and its contractors understand that construction of neutron producing reactors is not intended as a step in the super program.

PART II

Super Bombs

The General Advisory Committee has considered at great length the question of whether to pursue with high priority the development of the super bomb. No member of the Committee was willing to endorse this proposal. The reasons for our views leading to this conclusion stem in large part from the technical nature of the super and of the work necessary to establish it as a weapon. We therefore here transmit "an elementary" account of these matters.

The basic principle of design of the super bomb is the ignition of the thermo-nuclear DD reaction by the use of a fission bomb, and of high temperatures, pressure, and neutron densities which accompany it. In overwhelming probability, tritium is required as an intermediary, more easily ignited than the deuterium itself and, in turn, capable of igniting the deuterium. The steps which need to be taken if the super bomb is to become a reality include:

(1) The provision of tritium in amounts perhaps of several [deleted] per unit.
(2) Further theoretical studies and criticisms aimed at reducing the very great uncertainties still inherent in the behavior of this weapon under extreme conditions of temperature, pressure and flow.
(3) The engineering of designs which may on theoretical grounds appear hopeful, particularly with regard to the [deleted] problems presented.
(4) Carefully instrumented test programs to determine whether the deuterium-tritium mixture will be ignited by the fission bomb, . . .

It is notable that there appears to be no experimental approach short of actual test which will substantially add to our conviction that a given model will or will not work, and it is also notable that because of the unsymmetric and extremely unfamiliar conditions obtaining, some considerable doubt will surely remain as to the soundness of theoretical anticipation. Thus we are faced with a development which cannot be carried to the point of conviction without the actual construction and demonstration of the essential elements of the weapon in question. This does not mean that further theoretical studies would be without avail. It does mean that they could not be decisive. A final point that needs to be stressed is that many tests may be required before a workable model has been evolved or before it has been established beyond reasonable doubt that no such model can be evolved. Although we are not able to give a specific probability rating for any given model, we believe that an imaginative and concerted attack on the problem has a better than even chance of producing the weapon within five years.

A second characteristic of the super bomb is that once the problem of initiation has been solved, there is no limit to the explosive power of the bomb itself except that imposed by requirements of delivery. This is because one can continue to add deuterium—an essentially cheap material—to make larger and larger explosions, the energy release and radioactive products of which are both proportional to the amount of deuterium itself. Taking into account the probable limitations of carriers likely to be available for the delivery of such a weapon, it has generally been estimated that the weapon would have an explosive effect some hundreds of times that of present fission bombs. This would correspond to a damage area of the order of hundreds of square miles, to thermal radiation effects extending over a comparable area, and to very grave contamination problems which can easily be made more acute, and may possibly be rendered less acute, by surrounding the deuterium with uranium or other material. It needs to be borne in mind that for delivery by ship, submarine or other such carrier, the limitations here outlined no longer apply and that the weapon is from a technical point of view without limitation with regard to the damage that it can inflict.

It is clear that the use of this weapon would bring about the destruction of innumerable human lives; it is not a weapon which can be used exclusively for the destruction of material installations of military or semi-military purposes. Its use therefore carries much further than the atomic bomb itself the policy of exterminating civilian populations. It is of course true that super bombs which are not as big as those here contemplated could be made, provided the initiating mechanism works. In this case, however, there appears to be no chance of their being an economical alternative to the fission weapons themselves. It is clearly impossible with the vagueness of design and the uncertainty as to performance as we have them at present to give anything like a cost estimate of the super. If one uses the strict criteria of damage area per dollar and if one accepts the limitations on air carrier capacity likely to obtain in the years immediately ahead, it appears uncertain to us whether the super will be cheaper or more expensive than the fission bomb.

PART III

Although the members of the Advisory Committee are not unanimous in their proposals as to what should be done with regard to the super bomb, there are certain elements of unanimity among us. We all hope that by one means or another, the development of these weapons can be avoided. We are all reluctant to see the United States take the initiative in precipitating this development. We are all agreed that it would be wrong at the present moment to commit ourselves to an all-out effort toward its development.

We are somewhat divided as to the nature of the commitment not to develop the weapon. The majority feel that this should be an unqualified commitment. Others feel that it should be made conditional on the response of the Soviet government to a proposal to renounce such development. The Committee recommends that enough be declassified about the super bomb so that a public statement of policy can be made at this time. Such a statement might in our opinion point to the use of deuterium as the principal source of energy. It need not discuss initiating mechanisms nor the role which we believe tritium will play. It should explain that the weapon cannot be explored without developing it and proof-firing it. In one form or another the statement should express our desire not to make this development. It should explain the scale and general nature of the destruction which its use would entail. It should make clear that there are no known or foreseen nonmilitary applications of this development. The separate views of the members of the Committee are attached to this report for your use.

J. R. Oppenheimer

October 30, 1949

We have been asked by the Commission whether or not they should immediately initiate an "all-out" effort to develop a weapon whose energy release is 100 to 1000 times greater and whose destructive power in terms of area of damage is 20 to 100 times greater than those of the present atomic bomb. We recommend strongly against such action.

We base our recommendation on our belief that the extreme dangers to mankind inherent in the proposal wholly outweigh any military advantage that could come from this development. Let it be clearly realized that this is a super weapon; it is in a totally different category from an atomic bomb. The reason for developing such super bombs would be to have the capacity to devastate a vast area with a single bomb. Its use would involve a decision to slaughter a vast number of civilians. We are alarmed as to the possible global effects of the radioactivity generated by the explosion of a few super bombs of conceivable magnitude. If super bombs will work at all, there is no inherent limit in the destructive power that may be attained with them. Therefore, a super bomb might become a weapon of genocide.

The existence of such a weapon in our armory would have far-reaching effects on world opinion: reasonable people the world over would realize that the existence of a weapon of this type whose power of destruction is essentially unlimited represents a threat to the future of the human race which is intolerable. Thus we believe that the psychological effect of the weapon in our hands would be adverse to our interest.

We believe a super bomb should never be produced. Mankind would be far better off not to have a demonstration of the feasibility of such a weapon, until the present climate of world opinion changes.

It is by no means certain that the weapon can be developed at all and by no means certain that the Russians will produce one within a decade. To the argument that the Russians may succeed in developing this weapon, we would reply that our undertaking it will not prove a deterrent to them. Should they use the weapon against us, reprisals by our large stock of atomic bombs would be comparably effective to the use of a super.

In determining not to proceed to develop the super bomb, we see a unique opportunity of providing by example some limitations on the totality of war and thus of limiting the fear and arousing the hopes of mankind.

James B. Conant
Hartley Rowe
Cyril Stanley Smith
L. A. DuBridge
Oliver E. Buckley
J. R. Oppenheimer

October 30, 1949

An Opinion on the Development of the "super"

A decision on the proposal that an all-out effort be undertaken for the development of the "Super" cannot in our opinion be separated from considerations of broad national policy. A weapon like the "Super" is only an advantage when its energy release is from 100–1000 times greater than that of ordinary atomic bombs. The area of destruction therefore would run from 150 to approximately 1000 square miles or more.

Necessarily such a weapon goes far beyond any military objective and enters the range of very great natural catastrophes. By its very nature it cannot be confined to a military objective but becomes a weapon which in practical effect is almost one of genocide.

It is clear that the use of such a weapon cannot be justified on any ethical ground which gives a human being a certain individuality and dignity even if he happens to be a resident of an enemy country. It is evident to us that this would be the view of peoples in their countries. Its use would put the United States in a bad moral position relative to the peoples of the world.

Any postwar situation resulting from such a weapon would leave unresolvable enmities for generations. A desirable peace cannot come from such an inhuman application of force. The postwar problems would dwarf the problems which confront us at present.

The application of this weapon with the consequent great release of radioactivity would have results unforeseeable at present, but would certainly render large areas unfit for habitation for long periods of time.

The fact that no limits exist to the destructiveness of this weapon makes its very existence and the knowledge of its construction a danger to humanity as a whole. It is necessarily an evil thing considered in any light.

For these reasons we believe it important for the President of the United States to tell the American public, and the world, that we think it wrong on fundamental ethical principles to initiate a program of development of such a weapon. At the same time it would be appropriate to invite the nations of the world to join us in a solemn pledge not to proceed in the development or construction of weapons of this category. If such a pledge were accepted even without control machinery, it appears highly probable that an advanced stage of development leading to a test by another power could be detected by available physical means. Furthermore, we have in our possession, on our stockpile of atomic bombs, the means for adequate "military" retaliation for the production or use of "super."

E. Fermi
I. I. Rabi

30. Lewis Strauss to Harry S. Truman, November 25, 1949

25 November 1949

Dear Mr. President:

As you know, the thermonuclear (super) bomb was suggested by scientists working at Los Alamos during the war. The current consideration of the super bomb was precipitated, I believe, by a memorandum which I addressed to my fellow Commissioners following your announcement on September 23rd of an atomic explosion in Russia. I participated in the discussions which were antecedent to the letter to you from the Commission on November 9th, but did not join in the preparation of the letter as I was then on the Pacific Coast. It was my belief that a comprehensive recommendation should be provided for you, embodying the judgement of the Commission (in the areas where it is competent), together with the views of the Departments of State and Defense. My colleagues, however, felt that you would prefer to obtain these views separately.

Differences on the broad question of policy between my associates as individuals were included in the Commission's letter to you, and it was correctly stated that the views of Commissioner Dean and mine were in substantial accord on the main issue. It is proper, I believe, that I should state them on my own responsibility and in my own words.

I believe that the United States must be as completely armed as any possible enemy. From this, it follows that I believe it unwise to renounce, unilaterally, any weapon which an enemy can reasonably be expected to possess. I recommend that the President direct the Atomic Energy Commission to proceed with the development of the thermonuclear bomb, at highest priority subject only to the judgment of the Department of Defense as to its value as a weapon, and of the advice of the Department of State as to the diplomatic consequences of its unilateral renunciation or its possession. In the event that you may be interested, my reasoning is appended in a memorandum.

/s/ Lewis L. Strauss

25 November 1949

This is a memorandum to accompany a letter of even date to the President to supply the reasoning for my recommendation that he should direct the

Excerpt from *Men and Decisions* by Lewis Strauss. Copyright © 1962 by Lewis L. Strauss and the Lewis L. Strauss Literary Trust. Reprinted by permission of Doubleday & Company, Inc.

Atomic Energy Commission to proceed at highest priority with the development of the thermonuclear weapon.

Premises

(1) The production of such a weapon appears to be feasible (i.e., better than a 50–50 chance).
(2) Recent accomplishments by the Russians indicate that the production of a thermonuclear weapon is within their technical competence.
(3) A government of atheists is not likely to be dissuaded from producing the weapon on "moral" grounds. ("Reason and experience both forbid us to expect that national morality can prevail in exclusion of religious principle." G. Washington, September 17, 1796.)
(4) The possibility of producing the thermonuclear weapon was suggested more than six years ago, and considerable theoretical work has been done which may be known to the Soviets—the principle has certainly been known to them.
(5) The time in which the development of this weapon can be perfected is perhaps of the order of two years, so that a Russian enterprise started some years ago may be well along to completion.
(6) It is the historic policy of the United States not to have its forces less well armed than those of any other country (viz., the 5:5:3 naval ratio, etc. etc.).
(7) Unlike the atomic bomb which has certain limitations, the proposed weapon may be tactically employed against a mobilized army over an area of the size ordinarily occupied by such a force.
(8) The Commission's letter of November 9th to the President mentioned the "possibility that the radioactivity released by a small number (perhaps ten) of these bombs would pollute the earth's atmosphere to a dangerous extent." Studies requested by the Commission have since indicated that the number of such weapons necessary to pollute the earth's atmosphere would run into many hundreds. Atmospheric pollution is a consequence of present atomic bombs if used in quantity.

Conclusions

(1) The danger in the weapon does not reside in its physical nature but in human behavior. Its unilateral renunciation by the United States could very easily result in its unilateral possession by the Soviet Government. I am unable to see any satisfaction in that prospect.

(2)   The Atomic Energy Commission is competent to advise the President
with respect to the feasibility of making the weapon; its economy in
fissionable material as compared with atomic bombs; the possible time
factor involved; and a description of its characteristics compared to
atomic bombs. Judgment, however, as to its strategic or tactical importance for the armed forces should be furnished by the Department
of Defense, and views as to the effect on friendly nations or of unilateral renunciation of the weapon is a subject for the Department of
State. My opinion as an individual, however, based upon discussion
with military experts is to the effect that the weapon may be critically
useful against a large enemy force both as a weapon of offense and as a
defensive measure to prevent landings on our own shores.

(3)   I am impressed with the arguments which have been made to the
effect that this is a weapon of mass destruction on an unprecedented
scale. So, however, was the atomic bomb when it was first envisaged
and when the National Academy of Sciences in its report of November
6, 1941, referred to it as "of superlatively destructive power." Also on
June 16, 1945, the Scientific Panel of the Interim Committee on Nuclear Power, comprising some of the present members of the General
Advisory Committee, reported to the Secretary of War, "We believe
the subject of thermonuclear reactions among light nuclei is one of the
most important that needs study. There is a reasonable presumption
that with skillful research and development, fission bombs can be
used to initiate the reactions of deuterium, tritium, and possibly other
light nuclei. If this can be accomplished, the energy release of explosive units can be increased by a factor of 1000 or more over that of
presently contemplated fission bombs." This statement was preceded
by the recommendation, "Certainly we would wish to see work carried out on the problems mentioned below."

The General Advisory Committee to the Atomic Energy Commission, in its recent communication to the Commission recommending
against the development of the super bomb, noted that it "strongly
favors" the booster program, which is a program to increase the explosive power and hence the damage area and deadliness of atomic
bombs. These positions and those above appear not to be fully consistent and indicate that the scientific point of view is not unanimous.

(4)   Obviously the current atomic bomb as well as the proposed thermonuclear weapon are horrible to contemplate. All war is horrible. Until,
however, some means is found of eliminating war, I cannot agree with
those of my colleagues who feel that an announcement should be
made by the President to the effect that the development of the

thermonuclear weapon will not be undertaken by the United States at this time. This is because: (a) I do not think the statement will be credited in the Kremlin; (b) that when and if it should be decided subsequent to such a statement to proceed with the production of the thermonuclear bomb, it might in a delicate situation, be regarded as an affirmative statement of hostile intent; and (c) because primarily until disarmament is universal, our arsenal must be not less well equipped than with the most potent weapons that our technology can devise.

Recommendation

In sum, I believe that the president should direct the Atomic Energy Commission to proceed with all possible expedition to develop the thermonuclear weapon.

31. Statement by the President on the Hydrogen Bomb, January 31, 1950

It is part of my responsibility as Commander in Chief of the Armed Forces to see to it that our country is able to defend itself against any possible aggressor. Accordingly, I have directed the Atomic Energy Commission to continue its work on all forms of atomic weapons, including the so-called hydrogen or superbomb. Like all other work in the field of atomic weapons, it is being and will be carried forward on a basis consistent with the overall objectives of our program for peace and security.

This we shall continue to do until a satisfactory plan for international control of atomic energy is achieved. We shall also continue to examine all those factors that affect our program for peace and this country's security.

This text appeared in *Public Papers of the Presidents of the United States: Harry S. Truman, 1950* (Washington, D.C.: Government Printing Office, 1965), p. 138.

32. Hans Bethe, Comments on the History of the H-bomb, 1954

The H-bomb was suggested by Teller in 1942. Active work on it was pursued in the summer of 1942 by Oppenheimer, Teller, myself, and others (see Oppenheimer's testimony). The idea did not develop from Teller's "quiet work" at Los Alamos during the war.

When Los Alamos was started in Spring 1943, several groups of scientists were included who did work on this problem specifically. However, it was realized that this was a long-range project and that the main efforts of Los Alamos must be concentrated on making A-bombs. Teller, working on the H-bomb at Los Alamos, discovered a major difficulty (testimony by Oppenheimer). This discovery made it clear that it would be a very hard problem to make a "classical super" work, as this type of H-bomb was called. I shall refer to the classical super as Method A.

The work on thermonuclear weapons at Los Alamos never stopped. At this stage of the development, the main requirements were for theoretical work and for a few experimental physics measurements. Both of these types of work went ahead. On the basis of the monthly reports of the theoretical division of Los Alamos, it has been estimated that between 1946 and 1949 the work of that division was about equally divided between fission weapon design and problems related to thermonuclear weapons. (In this respect I was mistaken when testifying in the Oppenheimer case. I said then, from memory, that a relatively small fraction of the scientists of the division, though consisting of especially able men, were working on thermonuclear problems. Actually, the fraction was large.)

Two new methods of designing a thermonuclear weapon were invented (Methods B and C). Both inventions were due to Teller. Method B was invented in 1946, Method C in 1947. Method B was actively worked on by Richtmyer, Nordheim, and others. However, at that time, there seemed to be no way of putting Method B into practice, as Dr. Bradbury has mentioned in his statement to "The New Mexican." Teller himself wrote a most pessimistic report on the feasibility of this method in September 1947.

Method C is different from all the others in that thermonuclear reactions are used only in a minor way, for weapons of relatively small yield. This method seemed quite promising from the start, as early as the Summer of 1948 it was added to the devices to be tested in the Greenhouse tests.

This text appeared in *Los Alamos Science*, Fall 1982, pp. 43–53.

Theoretical work on the "classical super," Method A, proceeded continually, since this method was considered the most important of all thermonuclear devices. New plans for calculations were made frequently, mostly by consultation between Teller and the senior staff of the theoretical division. However, as Teller stated in 1946, "The required scientific effort is clearly much larger than that needed for the first fission weapon." In particular, the theoretical computations required were of such complication that they could not be handled in any reasonable time by any of the computing machines then available. Some greatly simplified calculations were done but it was realized that they left out many important factors and were therefore quite unreliable. Work was therefore concentrated on preparing full-scale calculations "for the time when adequate fast computing machines become available"—a sentence which recurs in many of the theoretical reports of this period. The plans for such a calculation on Method A were laid in September 1948, and the mathematical work was virtually completed by December 1949—all before the directive of President Truman— but it was not until mid-1952 that adequate computing machines finally became available, and by that time the most capable of them were fully engaged on the new and more promising proposal (Method D) discussed below.

When Dr. Teller and Admiral Strauss proposed in the Fall of 1949 to start a full-scale development of H-bombs, the method in their minds, as well as in the minds of the opponents of the program, was Method A. To accomplish Method A, two major problems had to be solved which I shall call Part 1 and Part 2. Part 1 seemed to be reasonably well in hand according to calculations made by Teller's group from 1944 to 1946 although nobody had been able to perform a really convincing calculation, as discussed in the paragraph above. Teller now believed that he had a solution for Part 2. In principle, the accomplishment of Part 2 had never been seriously in doubt, although the question of whether or not any particular device would behave in the way required could not be settled without experiment.

The Greenhouse thermonuclear experiment was designed to test Part 2. After President Truman made the decision to go ahead with a full-scale thermonuclear program, Los Alamos made plans to add to the Greenhouse test series an experiment intended to test a particular proposal relating to Part 2. Teller played a large part in the specification of this device, and as it turned out it behaved very well. However, as on previous occasions, Teller did not do so well in directing the detailed theoretical work of his group. Only as late as January 1951, a month or so before the test device had to be shipped to the Pacific, was the full theoretical prediction of the (probably successful) behavior of the device available. But even while complete theoretical proof was lacking, most of us connected with the work at Los

Alamos were confident that the Greenhouse experiment would work. As far as I could make out, at a meeting at Los Alamos in October 1950 which I attended as a guest, this was also the opinion of the GAC including Dr. Oppenheimer. Shepley and Blair instead report on page 116 that Dr. Oppenheimer expected the device to fail. (The correct story on Oppenheimer's attitude will be discussed below.)

A very large fraction of the members of the Los Alamos Laboratory, not just a "small handful of his" (Teller's) "associates" were extremely busy from Spring 1950 to Spring 1951 with the preparation of Teller's thermonuclear experiment. They did this in addition to preparing the Nevada tests of early 1951.

The major feature of the year 1950 was, however, the discovery that Part 1 of Method A was by no means under control. While Teller and most of the Los Alamos Laboratory were busy preparing the Greenhouse test, a number of persons in the theoretical division had continued to consider the various problems posed by Part 1. In particular, Dr. Ulam on his own initiative had decided to check the feasibility of aspects of Part 1 without the aid of high-speed computing equipment. He, and Dr. Everett who assisted him, soon found that the calculations of Teller's group of 1946 were wrong. Ulam's calculations showed that an extraordinarily large amount of tritium would be necessary. In the Summer of 1950 further calculations by Ulam and Fermi showed further difficulties with Part 1.

That Ulam's calculations had to be done at all was proof that the H-bomb project was not ready for a "crash" program when Teller first advocated such a program in the Fall of 1949. Nobody will blame Teller because the calculations of 1946 were wrong, especially because adequate computing machines were not then available. But he was blamed at Los Alamos for leading the Laboratory, and indeed the whole country, into an adventurous program on the basis of calculations which he himself must have known to have been very incomplete. The technical skepticism of the GAC on the other hand had turned out to be far more justified than the GAC itself had dreamed in October 1949.

We can now appreciate better the attitude of the GAC, and indeed of most of the members of Los Alamos, to the Greenhouse thermonuclear test. They did not expect it to fail, but they considered it as irrelevant because there appeared to be no solution to Part 1 of the problem. The correct description of this attitude is given by Oppenheimer in his own testimony.

The lack of a solid theoretical foundation was the only reason why the Los Alamos work might have seemed to some to have gotten off to a slow start in 1950. Purely theoretical work may seem slow in a project intended to develop "hardware," but there was simply no basis for building hardware until the theory had been clarified. As far as the mental attitude of Los Alamos in early 1950, it was almost the exact opposite of that described by Shepley and Blair. I visited Los Alamos around April 1, 1950 and tried to defend the point of view of the GAC in their decision of October 1949. I encountered almost universal hostility. The entire Laboratory seemed enthusiastic about the project and was working at high speed. That they continued to work with full energy on Teller's Greenhouse test, after Ulam's calculations had made the success of the whole program very doubtful, shows how far they were willing to go in following Teller's lead.

Teller himself was desperate between October 1950 and January 1951. He proposed a number of complicated schemes to save Method A, none of which seemed to show much promise. It was evident that he did not know of any solution. In spite of this, he urged that the Laboratory be put essentially at his disposal for another year or more after the Greenhouse test, at which time there should then be another test on some device or other. After the failure of the major part of his program in 1950, it would have been folly of the Los Alamos Laboratory to trust Teller's judgment, at least until he could present a definite idea which showed practical promise. This attitude was strongly held by most of those on the permanent staff of the Laboratory who were responsible for its operation. As might be expected, the many discussions of aspects of this situation bred considerable emotion.

Between January and May 1951, the "new concept" was developed. (This I shall call Method D.) In addition, it should be remembered that between January and May both tests in Nevada and the Greenhouse series of tests took place, and this required many senior members of the Laboratory to be at the test sites for prolonged periods of time and the attention of many others was engaged on study of results of these tests.

But what are the actual facts about this alleged delay in work on the new concept? In January Teller obviously did not know how to save the thermonuclear program. On March 9, 1951, according to Bradbury's press statement, Teller and Ulam published a paper which contained one-half of the new concept. As Bradbury has pointed out, Ulam as well as Teller should be given credit for this. Ulam, by the way, made his discovery while studying some aspects of fission weapons. This shows once more how the important ideas may not come from a straightforward attack on the main problem.

Within a month, the very important second half of the new concept occurred to Teller, and was given preliminary checks by de Hoffman. This immediately became the main focus of attention of the thermonuclear design program.

It is worth noting that the entire new concept was developed before the thermonuclear Greenhouse test which took place on May 8, 1951. The literature is full of statements that the success of Greenhouse was the direct cause of the new concept. This is historically false. Teller may have been influenced by thinking about the Greenhouse design when developing the new concept, but the success of Greenhouse (which was anticipated) had no influence on either the creation of the new concept, or on its quick adoption by the Laboratory or later by the GAC. The new concept stood on its own.

As early as the end of May 1951, I received from the Associate Director of Los Alamos a detailed proposal for the future program of the Laboratory in which Teller's new concept figured most prominently. By early June, when I visited Los Alamos for two weeks, everybody in the theoretical division was talking about the new concept.

Not only was the acceptance of the new concept not slow; but the realization of the development was a sensationally rapid accomplishment, in the same class as the achievement of Los Alamos during the war.

It is difficult to describe to a non-scientist the novelty of the new concept. It was entirely unexpected from the previous development. It was also not anticipated by Teller, as witness his despair immediately preceding the new concept. I believe that this very despair stimulated him to an invention that even he might not have made under calmer conditions. The new concept was to me, who had been rather closely associated with the program, about as surprising as the discovery of fission had been to physicists in 1939. Before 1939 scientists had a vague idea that it might be possible to release nuclear energy but nobody could think even remotely of a way to do it. If physicists had tried to discover a way to release nuclear energy before 1939, they would have worked on anything else rather than the field which finally led to the discovery of fission, namely radiochemistry. At that time, concentrated work on any "likely" way of releasing nuclear energy would have led nowhere. Similarly, concentrated work on Method A would never have led to Method D. The Greenhouse test had a vague connection with Method D but one that nobody, including Teller, could have foreseen or did foresee when that test was planned. By a misappraisal of the facts many persons not closely connected with the development have concluded that the scientists who had shown good judgment concerning the technical feasibility of Method A were now suddenly proved wrong, whereas Teller, who had been wrong in interpreting his own calculations, was suddenly right. The fact was that the new concept had created an entirely new technical situation. Such miracles incidentally do happen occasionally in scientific history but it would be folly to count on their occurrence. One of the dangerous consequences of the H-bomb history may well be that government administrators, and perhaps some scientists, too, will imagine that similar miracles should be expected in other developments.

Before the end of the Summer of 1951, the Los Alamos Laboratory was putting full force behind attempts to realize the new concept. However, the continued friction of 1950 and early 1951 had strained a number of personal relations between Teller and others at Los Alamos. In addition, Teller insisted on an earlier test date than the Laboratory deemed possible. There was further disagreement between Teller and Bradbury on personalities, in particular on the person who was to direct the actual development of hardware. Bradbury had great experience in administrative matters like these. Teller had no experience and had in the past shown no talent for administration. He had given countless examples of not completing the work he had started; he was inclined to inject constantly new modifications into an already going program which becomes intolerable in an engineering development beyond a certain stage; and he had shown poor technical judgment. Everybody recognizes that Teller more than anyone else contributed ideas at every stage of the H-bomb program, and this fact should never be obscured. However, as an article in "Life" of September 6, 1954, clearly portrays, nine out of ten of Teller's ideas are useless. He needs men with more judgment, even if they be less gifted, to select the tenth idea which often is a stroke of genius.

It has been loosely said that the people at Los Alamos couldn't "get along" with Teller and it might be worthwhile to clarify this point. Both during the difficulties of the wartime period and again in 1951, Teller was on excellent terms with the vast majority of the scientists at Los Alamos with whom he came in contact in the course of the technical work. On both occasions, however, friction arose between him and some of those responsible for the organization and operation of the Laboratory. In each case, Teller, who was essentially alone in his opinion, was convinced that things were hopelessly bad and that nothing would go right unless things were arranged quite differently. In each case, the Laboratory accomplished its mission with distinction. In September 1951, when the program for a specific test of the new concept was being planned, Teller was strongly urged to take the responsibility for directing the theoretical work on the design of Mike. But he felt sure the test date should be a few months earlier; he didn't like some of the people with whom he would have to work; he was convinced they weren't up to the job; the Laboratory was not organized properly and didn't have the right people. Teller decided to leave and left. The Mike shot went off exactly on schedule and was a full success.

It took much more than the idea of the new concept to design Mike.

Major difficulties occurred in the theoretical design in early 1952, which happened to be a period when I was again at Los Alamos. They were all solved by the splendid group of scientists at Los Alamos.

At this time more than one-half of all the development work of the Los Alamos Laboratory went into thermonuclear weapons and into the preparation of the Mike test in particular. All but a small percentage of the theoretical division were thinking about this subject. In addition, there was a group of theorists working in Princeton under the direction of Professor John A. Wheeler in collaboration with the theoretical group at Los Alamos. Shepley and Blair, however, have to say of this period (on page 141) "Progress on the thermonuclear program still lagged."

Teller "helped" at this time by intensive agitation against Los Alamos and for a second laboratory. This agitation was very disturbing to the few leading scientists at Los Alamos who knew about it. Much precious time was spent in trying to counteract Teller's agitation by bringing the true picture to Washington. I myself wrote a history of the thermonuclear development to Chairman Dean of the AEC which was mentioned in the Oppenheimer testimony. This loss of time could be ill afforded at a time when the technical preparations for Mike were in a crisis.

Nevertheless, the theoretical design of Mike was completed by June 1952 in good time to make the device ready for test on November 1. Not only this, but, in the same period, much work was done leading to the conceptual design of the devices which were later tested in the Castle series in the Spring 1954. The approximate date for the Castle tests was also set at that time, and it was planned then that it should lead to a deliverable H-bomb if the experimental Mike shot was successful. It is necessary always to plan approximately two years ahead. Between Summer 1952 and Spring 1954, theoretical calculations on the proposed thermonuclear weapons proceeded; they were followed and in some areas paralleled by mechanical design of the actual device and finally followed by manufacture of the "hardware".

In July 1952, the new laboratory at Livermore was officially established by the AEC. Its existence did not, and in fact could not, accelerate the Los Alamos work because in all essentials the work for Castle had been planned before Livermore was established. In August 1952 an additional device was conceived at Los Alamos which might possibly have been slightly influenced by ideas then beginning to be considered at Livermore. In addition, Los Alamos decided to make a few experimental small-scale shots in Nevada in the Spring of 1953, and this program may have been slightly stimulated by the existence of Livermore. Livermore did assist in the observation of the performance of some of the devices tested at Castle.

Concerning the performance of Livermore's own designs, I will only quote the statement of Dr. Bradbury to the press which says, "Every successful thermonuclear weapon tested so far" (1954) "has been developed by the Los Alamos Laboratory."[1] This statement has not been contradicted.

Was the H-bomb Necessary?

Although the GAC were seeking a solution rather than offering one, the proposal of its minority still seems worthwhile, even as seen from today's (1954) viewpoint. The proposal was to enter negotiations with Russia with the aim that both countries undertake an obligation not to develop the H-bomb. If such an agreement could have been reached and had been kept, it would have gone far to avoid the peril in which the world now stands. At that time neither we nor the Russians presumably knew whether an H-bomb could be made. In this blissful state of ignorance we might have remained for a long time to come. Since the technical program was a very difficult one it could never be accomplished without a major effort. It is possible, perhaps likely, that the Russians would have refused to enter an agreement on this matter. If they had done so, this refusal would have been a great propaganda asset for us in the international field and would in addition have gone far to persuade the scientists of this country to cooperate in the H-bomb program with enthusiasm.

Many people will argue that the Russians might have accepted such an agreement, but then broken it. I do not believe so. Thermonuclear weapons are so complicated that nobody will be confident that he has the correct solution before he has tested such a device. But it is well known that any test of a bomb of such high yield is immediately detected. Therefore, without any inspection, each side would know immediately if the other side had broken the agreement.

It is difficult to tell whether or not the Russians would have developed the H-bomb independently of us. I am not sure what would have happened if we had followed the recommendations of the GAC majority and had merely announced that for such and such reasons, we would refrain from developing the H-bomb. Once we announced that we would go ahead, the Russians clearly had no choice but to do the same. In the field of atomic weapons, we have called the tune since the end of the war, both

---

1. (*Note added in 1982:* In the intervening 28 years, Livermore has contributed greatly to nuclear weapons development. Some weapons programs are assigned to Livermore, some to Los Alamos, and the talents of the two laboratories complement each other.)

in quality and in quantity. Russia has to follow the tune or be a second-class power.

In summary I still believe that the development of the H-bomb is a calamity. I still believe that it was necessary to make a pause before the decision and to consider this irrevocable step most carefully. I still believe that the possibility of an agreement with Russia not to develop the bomb should have been explored. But once the decision was made to go ahead with the program, and once there was a sound technical program, I cooperated with it to the best of my ability. I did and still do this because it seems to me that once one is engaged in a race, one clearly must endeavor to win it. But one can try to forestall the race itself.

V

The Oppenheimer Case

During the winter of 1953–54, in the wake of the Korean War and several Soviet espionage cases involving atomic energy (notably the British physicist Klaus Fuchs and the Americans Ethel and Julius Rosenberg), the Cold War meant keeping atomic secrets from the Soviet Union. Both Moscow and Washington were directing major projects on developing a thermonuclear weapon as part of an escalating arms race. Los Alamos, together with the new Livermore national laboratory in California, was testing both atomic and hydrogen devices in the Nevada desert and on Pacific islands. As the power of the new weapons increased from kilotons to megatons in yield, and the size of deliverable weapons became ever smaller, many scientists had second thoughts about the power of the atom.

Among those scientists was J. Robert Oppenheimer, scientific director of the Manhattan Project and father of the atomic bomb, a man whose gaunt appearance and porkpie hat were as familiar to most Americans as a Norman Rockwell painting or Ike's grin. To the atomic scientists Oppie was a kind of father figure, a godlike administrator with a brilliant and incisive mind whose interests ranged from Sanskrit and French literature to Marxism and driving fast cars. Educated at Göttingen and Harvard, Oppie in the 1930s had commuted between the University of California at Berkeley and the California Institute of Technology at Pasadena, where he mesmerized a generation of young graduate students with his lectures on nuclear physics and quantum mechanics. Through his wife and his brother Frank, Oppie became involved in a number of then modish left-wing groups and causes in the California labor movement. He contributed money, he made new friends, he hosted meetings. In this he was not unlike many intellectuals of the 1930s for whom the forces of evil were represented by fascism and anti-Semitism, not communism.

From 1941 on Oppenheimer was deeply involved in the top secret project to build an American atomic bomb. In 1942 Arthur Holly Compton named him to work with General Groves as chief scientific director of the Manhattan Project. Oppenheimer's initial work at Berkeley and his later move to Los Alamos also attracted the attention of Army security agents concerned about his left-wing past and Communist friends. Even as Los Alamos was being constructed on a remote New Mexico mesa in early 1943, Oppie was being tailed by FBI and Army security agents and his telephone wiretapped. In July 1943 General Groves obtained a security clearance for Oppenheimer over the objections of army security advisers, claiming that Oppie was absolutely essential for the project whatever his political past.

With the success of the Manhattan Project, Oppenheimer's early political indiscretions and contacts were forgotten. After 1945 he left Los Alamos for Princeton, New Jersey, where he directed the Institute for Advanced Study. He served on the General Advisory Committee of the Atomic Energy Commission, a key panel of scientists, and became a familiar public speaker, author, and culture hero. To the public Oppie was a genius of a particular sort, an academic and intellectual who had unleashed the greatest power on earth—atomic energy—and struck a Faustian bargain with political and military authority.

In October 1949 Oppenheimer and the General Advisory Committee recommended against a crash program to build the hydrogen bomb. President Harry S. Truman, under pressure from Edward Teller and other scientific and military advisers, rejected that recommendation within months and decided to initiate the project. Despite his sense that the hydrogen bomb made no military sense and was essentially a weapon of genocide, Oppie participated in its early development for several years as a consultant to the Atomic Energy Commission. His opposition to it seemed forgotten. It was not.

In the winter of 1953–54, during the period of political paranoia associated with Senator Joseph McCarthy (R-Wis), Oppenheimer was surprised to receive a letter from Kenneth Nichols of the Atomic Energy Commission informing him that his security clearance would not be renewed. The letter included a number of accusations, innuendos, and charges relating to Oppie's early left-wing associations in California in the 1930s and 1940s and to his more recent opposition to the hydrogen bomb. Nichols informed Oppenheimer that if he wished he could request a hearing to defend himself against these charges that he was a security risk because of his politics and his ethics.

After writing a letter of response to Nichols, in April 1954 Oppenheimer confronted an Atomic Energy Commission three-man board of inquiry headed by Gordon Gray. His appearance promptly became a Gallilean drama of scientific truth and conscience pitted against irrational power and authority. The testimony of the witnesses, who included Groves, Teller, and most of the leading scientists in the Manhattan Project, largely defended Oppenheimer's character, patriotism, loyalty, and sense of integrity. Nevertheless, the Gray Board found Oppenheimer to be a security risk and upheld the AEC.

The result was a public outcry and a cathartic debate within the entire scientific community about Oppenheimer's guilt or innocence regarding vague charges. The publication by the government of the testimony and charges only confirmed the deep division between the larger purposes of scientific inquiry and military power in the name of national security. Although many scientists rallied to Oppie's defense, they were not free to disclose classified details of either fission or fusion bomb development, details that could have shown that any delay in the hydrogen bomb project was due not to Oppenheimer's politics but to Edward Teller's scientific mistakes. Teller himself had testified against Oppenheimer and had been ostracized by many in the scientific community; only in 1982 did it become known, in the classified 1954 article by Hans Bethe included above, that Teller himself was partly responsible for the delays in hydrogen bomb development.

Oppenheimer died in 1967, only partly vindicated by his government that had demanded his brilliant talents for so long and then discarded him as a scapegoat no longer necessary for nuclear weapons research.

The transcript of the Oppenheimer hearing is available from the U.S. government as In The Matter of J. Robert Oppenheimer: Transcript of Hearing Before Personnel Security Board, Washington, D.C., April 12, 1954, Through May 6, 1954 (Washington, D.C.: Government Printing Office, 1954; paperback ed. Cambridge, Mass.: MIT Press, 1971). On the case itself, see Philip M. Stern, with Harold P. Green, The Oppenheimer Case: Security on Trial (New York: Harper & Row, 1969). The best treatment of Oppenheimer's early years may be found in Alice Kimball Smith and Charles Weiner, eds., Robert Oppenheimer: Letters and Recollections (Cambridge, Mass.: Harvard University Press, 1980). A more recent study is James W. Kunetka, Oppenheimer: The Years of Risk (Englewood Cliffs, N.J.: Prentice-Hall, 1982).

33. K. D. Nichols, General Manager, AEC, to J. R. Oppenheimer, December 23, 1953

Section 10 of the Atomic Energy Act of 1946 places upon the Atomic Energy Commission the responsibility for assuring that individuals are employed by the Commission only when such employment will not endanger the common defense and security. In addition, Executive Order 10450 of April 27, 1953, requires the suspension of employment of any individual where there exists information indicating that his employment may not be clearly consistent with the interests of the national security.

As a result of additional investigation as to your character, associations, and loyalty, and review of your personnel security file in the light of the requirements of the Atomic Energy Act and the requirements of Executive Order 10450, there has developed considerable question whether your continued employment on Atomic Energy Commission work will endanger the common defense and security and whether such continued employment is clearly consistent with the interests of the national security. This letter is to advise you of the steps which you may take to assist in the resolution of this question.

The substance of the information which raises the question concerning your eligibility for employment on Atomic Energy Commission work is as follows:

It was reported that in 1940 you were listed as a sponsor of the Friends of the Chinese People, an organization which was characterized in 1944 by the House Committee on Un-American Activities as a Communist-front organization. It was further reported that in 1940 your name was included on a letterhead of the American Committee for Democratic and Intellectual Freedom as a member of its national executive committee. The American Committee for Democracy and Intellectual Freedom was characterized in 1942 by the House Committee on Un-American Activities as a Communist front which defended Communist teachers, and in 1943 it was characterized as subversive and un-American by a special subcommittee of the House Committee on Appropriations. It was further reported that in 1938 you were a member of the Western Council of the Consumers Union. The Consumers Union was cited in 1944 by the House Committee on Un-American Activities as a Communist-front headed by the Communist Ar-

This document appeared in *In the Matter of J. Robert Oppenheimer* (Washington, D.C.: Government Printing Office, 1954), pp. 3–7.

thur Kallet. It was further reported that you stated in 1943 that you were not a Communist, but had probably belonged to every Communist front organization on the west coast and had signed many petitions in which Communists were interested.

It was reported that in 1943 and previously you were intimately associated with Dr. Jean Tatlock, a member of the Communist Party in San Francisco, and that Dr. Tatlock was partially responsible for your association with Communist-front groups.

It was reported that your wife, Katherine Puening Oppenheimer, was formerly the wife of Joseph Dallet, a member of the Communist Party, who was killed in Spain in 1937 fighting for the Spanish Republican Army. It was further reported that during the period of her association with Joseph Dallet, your wife became a member of the Communist Party. The Communist Party has been designated by the Attorney General as a subversive organization which seeks to alter the form of Government of the United States by unconstitutional means, within the purview of Executive Order 9835 and Executive Order 10450. . . .

It was reported that you have associated with members and officials of the Communist Party including Isaac Folkoff, Steve Nelson, Rudy Lambert, Kenneth May, Jack Manley, and Thomas Addis.

It was reported that you were a subscriber to the *Daily People's World*, a west coast Communist newspaper, in 1941 and 1942. . . .

It was reported that prior to March 1, 1943, possibly 3 months prior, Peter Ivanov, secretary of the Soviet consulate, San Francisco, approached George Charles Eltenton for the purpose of obtaining information regarding work being done at the Radiation Laboratory for the use of Soviet scientists; that George Charles Eltenton subsequently requested Haakon Chevalier to approach you concerning this matter; that Haakon Chevalier thereupon approached you, either directly or through your brother, Frank Friedman Oppenheimer, in connection with this matter; and that Haakon Chevalier finally advised George Charles Eltenton that there was no chance whatsoever of obtaining the information. It was further reported that you did not report this episode to the appropriate authorities until several months after its occurrence; that when you initially discussed this matter with the appropriate authorities on August 26, 1943, you did not identify yourself as the person who had been approached, and you refused to identify Haakon Chevalier as the individual who made the approach on behalf of George Charles Eltenton; and that it was not until several months later, when you were ordered by a superior to do so, that you so identified Haakon Chevalier. It was further reported that upon your return to Berkeley following your separation from the Los Alamos project, you were visited by the Chevaliers on several occasions; and that your wife was in contact with Haakon and Barbara Chevalier in 1946 and 1947.

It was reported that in 1945 you expressed the view that "there is a reasonable possibility that it (the hydrogen bomb) can be made," but that the feasibility of the hydrogen bomb did not appear, on theoretical grounds, as certain as the fission bomb appeared certain, on theoretical grounds, when the Los Alamos Laboratory was started; and that in the autumn of 1949 the General Advisory Committee expressed the view that "an imaginative and concerted attack on the problem has a better than even chance of producing the weapon within 5 years." It was further reported that in the autumn of 1949 and subsequently, you strongly opposed the development of the hydrogen bomb; (1) on moral grounds, (2) by claiming that it was not feasible, (3) by claiming that there were insufficient facilities and scientific personnel to carry on the development and (4) that it was not politically desirable. It was further reported that even after it was determined, as a matter of national policy, to proceed with development of a hydrogen bomb, you continued to oppose the project and declined to cooperate fully in the project. It was further reported that you departed from your proper role as an adviser to the Commission by causing the distribution separately and in private, to top personnel at Los Alamos of the majority and minority reports of the General Advisory Committee on development of the hydrogen bomb for the purpose of trying to turn such top personnel against the development of the hydrogen bomb. It was further reported that you were instrumental in persuading other outstanding scientists not to work on the hydrogen-bomb project, and that the opposition to the hydrogen bomb, of which you are the most experienced, most powerful, and most effective member, has definitely slowed down its development.

In view of your access to highly sensitive classified information, and in view of these allegations which, until disproved, raise questions as to your veracity, conduct and even your loyalty, the Commission has no other recourse, in discharge of its obligations to protect the common defense and security, but to suspend your clearance until the matter has been resolved. Accordingly, your employment on Atomic Energy Commission work and your eligibility for access to restricted data are hereby suspended, effective immediately, pending final determination of this matter.

To assist in the resolution of this matter, you have the privilege of appearing before an Atomic Energy Commission personnel security board. To avail yourself of the privileges afforded you under the Atomic Energy Commission hearing procedures, you must, within 30 days following receipt of this letter, submit to me, in writing, your reply to the information outlined above and request the opportunity of appearing before the personnel security board. Should you signify your desire to appear before the board, you will be notified of the composition of the board and may challenge any member of it for cause. Such challenge should be submitted within 72 hours of the receipt of notice of composition of the board. . . .

If a written response is not received from you within 30 days it will be assumed that you do not wish to submit any explanation for further consideration. In that event, or should you not advise me in writing of your desire to appear before the personnel security board, a determination in your case will be made by me on the basis of the existing record. . . .

Very truly yours,

K. D. Nichols, General Manager.

---

34. J. R. Oppenheimer to K. D. Nichols, March 4, 1954

In the spring of 1936, I had been introduced by friends to Jean Tatlock, the daughter of a noted professor of English at the university; and in the autumn, I began to court her, and we grew close to each other. We were at least twice close enough to marriage to think of ourselves as engaged. Between 1939 and her death in 1944 I saw her very rarely. She told me about her Communist Party memberships; they were on again, off again affairs, and never seemed to provide for her what she was seeking. I do not believe that her interests were really political. She loved this country and its people and its life. She was, as it turned out, a friend of many fellow travelers and Communists, with a number of whom I was later to become acquainted.

I should not give the impression that it was wholly because of Jean Tatlock that I made leftwing friends, or felt sympathy for causes which hitherto would have seemed so remote from me, like the Loyalist cause in Spain, and the organization of migratory workers. I have mentioned some of the other contributing causes. I liked the new sense of companionship, and at the time felt that I was coming to be part of the life of my time and country.

In 1937, my father died; a little later, when I came into an inheritance,

This document appeared in *In the Matter of J. Robert Oppenheimer* (Washington, D.C.: Government Printing Office, 1954), pp. 8–11.

I made a will leaving this to the University of California for fellowships to graduate students.

This was the era of what the Communists then called the United Front, in which they joined with many non-Communist groups in support of humanitarian objectives. Many of these objectives engaged my interest. I contributed to the strike fund of one of the major strikes of Bridges' union; I subscribed to the *People's World;* I contributed to the various committees and organizations which were intended to help the Spanish Loyalist cause. I was invited to help establish the teacher's union, which included faculty and teaching assistants at the university, and school teachers of the East Bay. I was elected recording secretary. My connection with the teacher's union continued until some time in 1941, when we disbanded our chapter. . . .

My own views were also evolving. Although Sidney and Beatrice Webb's book on Russia, which I had read in 1936, and the talk that I heard at that time had predisposed me to make much of the economic progress and general level of welfare in Russia, and little of its political tyranny, my views on this were to change. I read about the purge trials, though not in full detail, and could never find a view of them which was not damning to the Soviet system. In 1938 I met three physicists who had actually lived in Russia in the thirties. All were eminent scientists, Placzek, Weisskopf, and Schein; and the first two have become close friends. What they reported seemed to me so solid, so unfanatical, so true, that it made a great impression; and it presented Russia, even when seen from their limited experience, as a land of purge and terror, of ludicrously bad management and of a long-suffering people. I need to make clear that this changing opinion of Russia, which was to be reinforced by the Nazi-Soviet Pact, and the behavior of the Soviet Union in Poland and in Finland, did not mean a sharp break for me with those who held to different views. At that time I did not fully understand—as in time I came to understand—how completely the Communist Party in this country was under the control of Russia. During and after the battle of France, however, and during the battle of England the next autumn, I found myself increasingly out of sympathy with the policy of disengagement and neutrality that the Communist press advocated. . . .

Because of these associations that I have described, and the contributions mentioned earlier, I might well have appeared at the time as quite close to the Communist Party—perhaps even to some people as belonging to it. As I have said, some of its declared objectives seemed to me desirable. But I never was a member of the Communist Party. I never accepted Communist dogma or theory; in fact, it never made sense to me. I had no clearly formulated political views. I hated tyranny and repression and every form of dictatorial control of thought. In most cases I did not in those days know who was and who was not a member of the Communist Party. No one ever asked me to join the Communist Party. . . .

In 1943 when I was alleged to have stated that "I knew several individuals then at Los Alamos who had been members of the Communist Party," I knew of only one; she was my wife, of whose disassociation from the party, and of whose integrity and loyalty to the United States I had no question. Later, in 1944 or 1945, my brother Frank, who had been cleared for work in Berkeley and at Oak Ridge, came to Los Alamos from Oak Ridge with official approval.

I knew of no attempt to obtain secret information at Los Alamos. Prior to my going there my friend Haakon Chevalier with his wife visited us on Eagle Hill, probably in early 1943. During the visit, he came into the kitchen and told me that George Eltenton had spoken to him of the possibility of transmitting technical information to Soviet scientists. I made some strong remark to the effect that this sounded terribly wrong to me. The discussion ended there. Nothing in our long standing friendship would have led me to believe that Chevalier was actually seeking information; and I was certain that he had no idea of the work on which I was engaged.

It has long been clear to me that I should have reported the incident at once. The events that led me to report it—which I doubt ever would have become known without my report—were unconnected with it. During the summer of 1943, Colonel Landsdale, the intelligence officer of the Manhattan District, came to Los Alamos and told me that he was worried about the security situation in Berkeley because of the activities of the Federation of Architects, Engineers, Chemists, and Technicians. This recalled to my mind that Eltenton was a member and probably a promoter of the FAECT. Shortly thereafter, I was in Berkeley and I told the security officer that Eltenton would bear watching. When asked why, I said that Eltenton had attempted, through intermediaries, to approach people on the project, though I mentioned neither myself nor Chevalier. Later, when General Groves urged me to give the details, I told him of my conversation with Chevalier. I still think of Chevalier as a friend. . . .

From the close of the war, when I returned to the west coast until finally in the spring of 1947 when I went to Princeton as the director of the Institute for Advanced Study, I was able to spend very little time at home and in teaching in California. In October 1945, at the request of Secretary of War Patterson, I had testified before the House Committee on Military Affairs in support of the May-Johnson bill, which I endorsed as an interim means of bringing about without delay the much needed transition from the wartime administration of the Manhattan District to postwar management of the atomic-energy enterprise. In December 1945, and later, I appeared at Senator McMahon's request in sessions of his Special Committee on Atomic Energy, which was considering legislation on the same subject. Under the chairmanship of Dr. Richard Tolman, I served on a committee set up by General Groves to consider classification policy on matters of atomic energy. For 2 months, early in 1946, I worked steadily as a member of a panel, the Board of Consultants to the Secretary of State's Committee on Atomic Energy, which, with the Secretary of State's Committee, prepared the so-called Acheson-Lilienthal report. After the publication of this report, I spoke publicly in support of it. A little later, when Mr. Baruch was appointed to represent the United States in the United Nations Atomic Energy Committee, I became one of the scientific consultants to Mr. Baruch, and his staff in preparation for and in the conduct of our efforts to gain support for the United State's plan. I continued as a consultant to General Osborn when he took over the effort.

At the end of 1946 I was appointed by the President as a member of the General Advisory Committee to the Atomic Energy Commission. At its first meeting I was elected Chairman, and was reelected until the expiration of my term in 1952. This was my principal assignment during these years as far as the atomic-energy program was concerned, and my principal preoccupation apart from academic work. . . .

The initial members of the General Advisory Committee were Conant, then president of Harvard, DuBridge, president of the California Institute of Technology, Fermi of the University of Chicago, Rabi of Columbia University, Rowe, vice president of the United Fruit Co., Seaborg of the University of California, Cyril Smith of the University of Chicago, and Worthington of the duPont Co. In 1948 Buckley, president of the Bell Telephone Laboratories, replaced Worthington; in the summer of 1950, Fermi, Rowe, and Seaborg were replaced by Libby of the University of Chicago, Murphree, president of Standard Oil Development Co., and Whitman of the Massachusetts Institute of Technology. Later Smith resigned and was succeeded by von Neumann of the Institute for Advance Study.

In these years from early 1947 to mid-1952 the Committee met some 30 times and transmitted perhaps as many reports to the Commission. Formulation of policy and the management of the vast atomic-energy enterprises were responsibilities vested in the Commission itself. The General Advisory Committee had the role, which was fixed for it by statute, to advise the Commission. In that capacity we gave the Commission our views on questions which the Commission put before us, brought to the

Commission's attention on our initiative technical matters of importance, and encouraged and supported the work of the several major installations of the Commission. . . .

The super itself had a long history of consideration beginning, as I have said, with our initial studies in 1942 before Los Alamos was established. It continued to be the subject of study and research at Los Alamos throughout the war. After the war, Los Alamos itself was inevitably handicapped pending the enactment of necessary legislation for the atomic energy enterprise. With the McMahon Act, the appointment of the Atomic Energy commission and the General Advisory Committee, we in the committee had occasion at our early meetings in 1947 as well as in 1948 to discuss the subject. In that period the General Advisory Committee pointed out the still extremely unclear status of the problem from the technical standpoint, and urged encouragement of Los Alamos' efforts which were then directed toward modest exploration of the super and of thermonuclear systems. No serious controversy arose about the super until the Soviet explosion of an atomic bomb in the autumn of 1949.

Shortly after that event, in October 1949, the Atomic Energy Commission called a special session of the General Advisory Committee and asked us to consider and advise on two related questions: First, whether in view of the Soviet success the Commission's program was adequate, and if not, in what way it should be altered or increased; second, whether a crash program for the development of the super should be a part of any new program. The committee considered both questions, consulting various officials from the civil and military branches of the executive departments who would have been concerned, and reached conclusions which were communicated in a report to the Atomic Energy Commission in October 1949.

This report, in response to the first question that had been put to us, recommended a great number of measures that the Commission should take to increase in many ways our overall potential in weapons.

As to the super itself, the General Advisory Committee stated its unanimous opposition to the initiation by the United States of a crash program of the kind we had been asked to advise on. The report of that meeting, and the Secretary's notes, reflect the reasons which moved us to this conclusion. The annexes, in particular, which dealt more with political and policy considerations—the report proper was essentially technical in character—indicated differences in the views of members of the committee. There were two annexes, one signed by Rabi and Fermi, the other by Conant, DuBridge, Smith, Rowe, Buckley and myself. (The ninth member of the committee, Seaborg, was abroad at the time.)

It would have been surprising if eight men considering a problem of extreme difficulty had each had precisely the same reasons for the conclusion in which we joined. But I think I am correct in asserting that the unanimous opposition we expressed to the crash program was based on the conviction, to which technical considerations as well as others contributed, that because of our overall situation at that time such a program might weaken rather than strengthen the position of the United States.

After the report was submitted to the Commission, it fell to me as chairman of the committee to explain our position on several occasions, once at a meeting of the Joint Congressional Committee on Atomic Energy. All this, however, took place prior to the decision by the President to proceed with the thermonuclear program.

This is the full story of my "opposition to the hydrogen bomb." It can be read in the records of the general transcript of my testimony before the joint congressional committee. It is a story which ended once and for all when in January 1950 the President announced his decision to proceed with the program. I never urged anyone not to work on the hydrogen bomb project. I never made or caused any distribution of the GAC reports except to the Commission itself. As always, it was the Commission's responsibility to determine further distribution.

In summary, in October 1949, I and the other members of the General Advisory Committee were asked questions by the Commission to which we had a duty to respond, and to which we did respond with our best judgment in the light of evidence then available to us. . . .

In this letter, I have written only of those limited parts of my history which appear relevant to the issue now before the Atomic Energy Commission. In order to preserve as much as possible the perspective of the story, I have dealt very briefly with many matters. I have had to deal briefly or not at all with instances in which my actions or views were adverse to Soviet or Communist interest, and of actions that testify to my devotion to freedom, or that have contributed to the vitality, influence and power of the United States.

In preparing this letter, I have reviewed two decades of my life. I have recalled instances where I acted unwisely. What I have hoped was, not that I could wholly avoid error, but that I might learn from it. What I have learned has, I think, made me more fit to serve my country.

Very truly yours,

J. Robert Oppenheimer

Princeton, N.J., March 4, 1954.

35. Testimony in the Matter of J. Robert Oppenheimer

a. General Leslie B. Groves.

Q. General, did your security officers on the project advise against the clearance of Dr. Oppenheimer?

A. Oh, I am sure that they did. I don't recall exactly. They certainly were not in favor of his clearance. I think a truer picture is to say that they reported that they could not and would not clear him.

Q. General, you were in the Army actively for how many years?

A. I don't know. 1916 to 1948, and of course raised in it, also.

Q. And you rose to the rank of lieutenant general?

A. That is right.

Q. During your entire Army career, I assume you were dealing with matters of security?

A. I would say I devoted about 5 percent of my time to security problems.

Q. You did become thoroughly familiar with security matters.

A. I think that I was very familiar with security matters.

Q. In fact, it could be said that you became something of an expert in it?

A. I am afraid that is correct.

Q. I believe you said that you became pretty familiar with the file of Dr. Oppenheimer?

A. I think I was thoroughly familiar with everything that was reported about Dr. Oppenheimer; and that included, as it did on every other matter of importance, personally reading the original evidence if there was any original evidence. In other words, I would read the reports of the interviews with people. In other words, I was not reading the conclusions of any security officer. The reason for that was that in this project there were so many things that the security officer would not know the significance of that I felt I had to do it myself. Of course, I have been criticized for doing all those things myself and not having a staff of any kind; but, after all, it did work, and I did live through it.

Q. General, in the light of your experience with security matters and in

This document appeared in *In the Matter of J. Robert Oppenheimer* (Washington, D.C.: Government Printing Office, 1954), pp. 170–71, 330–31, 356–57, 468–69, 710.

the light of your knowledge of the file pertaining to Dr. Oppenheimer, would you clear Dr. Oppenheimer today?

A. I think before answering that I would like to give my interpretation of what the Atomic Energy Act requires. I have it, but I never can find it as to just what it says. Maybe I can find it this time.

Q. Would you like me to show it?

A. I know it is very deeply concealed in the thing.

Q. Do you have the same copy?

A. I have the original act.

Q. It is on page 14, I think, where you will find it, General. You have the same pamphlet I have.

A. Thank you. That is it. The clause to which I am referring is this: It is the last of paragraph (b) (i) on page 14. It says:

"The Commission shall have determined that permitting such person to have access to restricted data will not endanger the common defense or security," and it mentions that the investigation should include the character, associations, and loyalty.

My interpretation of "endanger"—and I think it is important for me to make that if I am going to answer your question—is that it is a reasonable presumption that there might be a danger, not a remote possibility, a tortured interpretation of maybe there might be something, but that there is something that might do. Whether you say that is 5 percent or 10 percent or something of that order does not make any difference. It is not a case of proving that the man is a danger. It is a case of thinking, well, he might be a danger, and it is perfectly logical to presume that he would be, and that there is no consideration whatsoever to be given to any of his past performances or his general usefulness or, you might say, the imperative usefulness. I don't care how important the man is, if there is any possibility other than a tortured one that his associations or his loyalty or his character might endanger.

In this case I refer particularly to associations and not to the associations as they exist today but the past record of the associations. I would not clear Dr. Oppenheimer today if I were a member of the Commission on the basis of this interpretation.

If the interpretation is different, then I would have to stand on my interpretation of it.

b. Hans Bethe.

Q. Do you have any opinion, Dr. Bethe, on the question of whether there has been in fact any delay in the development and the perfection of thermonuclear weapons by the United States?

A. I do not think that there has been any delay. I will try to keep this unclassified. I can't promise that I can make myself fully clear on this.

Q. Try to, will you?

A. I will try. When President Truman decided to go ahead with the hydrogen bomb in January 1950, there was really no clear technical program that could be followed. This became even more evident later on when new calculations were made at Los Alamos, and when these new calculations showed that the basis for technical optimism which had existed in the fall of 1949 was very shaky, indeed. The plan which then existed for the making of a hydrogen bomb turned out to be less and less promising as time went on.

Q. What interval are you now speaking of?

A. I am speaking of the interval of from January 1950 to early 1951. It was a time when it would not have been possible by adding more people to make any more progress. The more people would have to do would have to be work on the things which turned out to be fruitful.

Finally there was a very brilliant discovery made by Dr. Teller. It was one of the discoveries for which you cannot plan, one of the discoveries like the discovery of the relativity theory, although I don't want to compare the two in importance. But something which is a stroke of genius, which does not occur in the normal development of ideas. But somebody has to suddenly have an inspiration. It was such an inspiration which Dr. Teller had which put the program on a sound basis.

Only after there was such a sound basis could one really talk of a technical program. Before that, it was essentially only speculation, essentially only just trying to do something without having really a direction in which to go. Now things changed very much. After this brilliant discovery there was a program.

Q. Dr. Bethe, if the board and Mr. Robb would permit me, I would like to ask you somewhat a hypothetical question. Would your attitude about work on the thermonuclear program in 1949 have differed if at that time there had been available this brilliant discovery or brilliant inspiration, whatever you call it, that didn't come to Teller until the spring of 1951?

A. It is very difficult to answer this.

Q. Don't answer it if you can't.

A. I believe it might have been different.

Q. Why?

A. I was hoping that it might be possible to prove that thermonuclear reactions were not feasible at all. I would have thought that the greatest security for the United States would have lain in the conclusive proof of the impossibility of a thermonuclear bomb. I must confess that this was the main motive which made me start work on thermonuclear reactions in the summer of 1950.

With the new [Teller-Ulam idea?] I think the situation changed because it was then clear, or almost clear—at least very likely—that thermonuclear weapons were indeed possible. If thermonuclear weapons were possible, I felt that we should have that first and as soon as possible. So I think my attitude might have been different.

Q. One final question, Dr. Bethe. I should have asked you this. I have referred you to the press statements and the article that you published in the late winter and spring of 1950, expressing critical views of the H-bomb program. Did you ever discuss those moves, that is to make such statements and write such articles, with Dr. Oppenheimer?

A. I never did. In fact, after the President's decision, he would never discuss any matters of policy with me. There had been in fact a directive from President Truman to the GAC not to discuss the reasons of the GAC or any of the procedures, and Dr. Oppenheimer held to this directive very strictly.

Q. Did you consult him about the article?

A. I don't think I consulted him at all about the article. I consulted him about the statement that we made. As far as I remember, he gave no opinion.

Q. On the basis of your association with him, your knowledge of him over these many years, would you care to express an opinion about Dr. Oppenheimer's loyalty to the United States, about his character, about his discretion in regard to matters of security?

A. I am certainly happy to do this. I have absolute faith in Dr. Oppenheimer's loyalty. I have always found that he had the best interests of the United States at heart. I have always found that if he differed from other people in his judgment, that it was because of a deeper thinking about the possible consequences of our action than the other people had. I believe that it is an expression of loyalty—of particular loyalty—if a person tries to go beyond the obvious and tries to make available his deeper insight, even in making unpopular suggestions, even in making suggestions which are not the obvious ones to make, are not those which a normal intellect might be led to make.

I have absolutely no question that he has served this country very long and very well. I think everybody agrees that his service in Los Alamos was one of the greatest services that were given to this country. I believe he has served equally well in the GAC in reestablishing the strength of our atomic weapons program in 1947. I have faith in him quite generally.

Q. You and he are good friends?

A. Yes.

Q. Would you expect him to place his loyalty to his country even above his loyalty to a friend?

A. I suppose so.

Mr. Marks. That is all.

c. George F. Kennan.

Q. As a result of your experience with Dr. Oppenheimer in the cases that you have reference to, what convictions, if any, did you form about him?

A. I formed the conviction that he was an immensely useful person in the councils of our Government, and I felt a great sense of gratitude that we had his help. I am able to say that in the course of all these contacts and deliberations within the Government I never observed anything in his conduct or his words that could possibly, it seemed to me, have indicated that he was animated by any other motives than a devotion to the interests of this country.

Q. Did you ever observe anything that would possibly have suggested to you that he was taking positions that the Russians would have liked?

A. No. I cannot say that I did in any way. After all, the whole purpose of these exercises was to do things which were in the interest of this country, not in the interests of the Soviet Union, at least not in the interests of the Soviet Union as their leaders saw it at that time. Anyone who collaborated sincerely and enthusiastically in the attempt to reach our objectives, which Dr. Oppenheimer did, obviously was not serving Soviet purposes in any way.

Q. Have you said that he contributed significantly to the results?

A. I have, sir.

Q. Mr. Kennan, is there any possibility in your mind that he was dissembling?

A. There is in my mind no possibility that Dr. Oppenheimer was dissembling.

Q. How do you know that? How can anybody know that?

A. I realize that is not an assertion that one could make with confidence about everyone. If I make it with regard to Dr. Oppenheimer it is because I feel and believe that after years of seeing him in various ways, not only there in Government, but later as an associate and a neighbor, and a friend at Princeton, I know his intellectual makeup and something of his personal makeup and I consider it really out of the question that any man could have participated as he did in these discussions, could have bared his thoughts to us time after time in the way that he did, could have thought those thoughts, so to speak, in our presence and have been at the same time dissembling.

I realize that is still not wholly the answer. The reason I feel it is out of the question that could have happened is that I believed him to have an intellect of such a nature that it would be impossible for him to speak dishonestly about any subject to which he had given his deliberate and careful and professional attention.

That is the view I hold of him. I have the greatest respect for Dr. Oppenheimer's mind. I think it is one of the great minds of this generation of Americans. A mind like that is not without its implications.

Q. Without its what?

A. Implications for a man's general personality. I think it would be actually the one thing probably in life that Dr. Oppenheimer could never do, that is to speak dishonestly about a subject which had really engaged the responsible attention of his intellect. My whole impression of him is that he is a man who when he turns his mind to something in an orderly and responsible way, examines it with the most extraordinary scrupulousness and fastidiousness of intellectual process.

I must say that I cannot conceive that in these deliberations in Government he could have been speaking disingenuously to us about these matters. I would suppose that you might just as well have asked Leonardo da Vinci to distort an anatomical drawing as that you should ask Robert Oppenheimer to speak responsibly to the sort of questions we were talking about, and speak dishonestly.

Q. Mr. Kennan, in saying what you have just said, are you saying it with an awareness of the background that Dr. Oppenheimer has, the general nature of which is reflected in the letter which General Nichols addressed to him, which is the genesis of these proceedings, and his response?

A. I am, sir.

Q. How do you reconcile these two things?

A. I do not think that they are necessarily inconsistent one with the other. People advance in life for one thing. I saw Dr. Oppenheimer at a phase of his life in which most of these matters in General Nichols' letter did not apply. It seems to me also that I was concerned or associated with him in the examination of problems which both he and I had accepted as problems of governmental responsibility before us, and I do not suppose that was the case with all the things that were mentioned in General Nichols' letter about his early views about politics and his early activities and his early associations.

I also think it quite possible for a person to be himself profoundly honest and yet to have associates and friends who may be misguided and misled and for who either at the time or in retrospect he may feel intensely sorry and concerned. I think most of us have had the experience of having known people at one time in our lives of whom we felt that way.

d. Isador Rabi.

Q. Dr. Rabi, Mr. Robb asked you whether you had spoken to Chairman Strauss in behalf of Dr. Oppenheimer. Did you mean to suggest in your reply—in your reply to him you said you did among other things—did you mean to suggest that you had done that at Dr. Oppenheimer's instigation?

A. No; I had no communication from Dr. Oppenheimer before these charges were filed, or since, except that I called him once to just say that I believed in him, with no further discussion.

Another time I called on him and his attorney at the suggestion of Mr. Strauss. I never hid my opinion from Mr. Strauss that I thought that this whole proceeding was a most unfortunate one.

Dr. Evans: What was that?

The Witness: That the suspension of the clearance of Dr. Oppenheimer was a very unfortunate thing and should not have been done. In other words, there he was; he is a consultant, and if you don't want to consult the guy, you don't consult him, period. Why you have to then proceed to suspend clearance and go through all this sort of thing, he is only there when called, and that is all there was to it. So it didn't seem to me the sort of thing that called for this kind of proceeding at all against a man who had accomplished what Dr. Oppenheimer has accomplished. There is a real positive record, the way I expressed it to a friend of mine. We have an A-bomb and a whole series of it, and what more do you want, mermaids? This is just a tremendous achievement. If the end of that road is this kind of hearing, which can't help but be humiliating, I thought it was a pretty bad show. I still think so.

By Mr. Marks:

Q. Dr. Rabi, in response to a question of the Chairman, the substance of which I believe was, was Dr. Oppenheimer unalterably opposed to the H-bomb development at the time of the October 1949 GAC meeting, I think you said in substance no, and then you added by way of explanation immediately thereafter the two annexes or whatever they were—

A. During the discussion.

Q. During the discussion he said he would be willing to sign either or both. Can you explain what you meant by that rather paradoxical statement?

A. No, I was just reporting a recollection.

Q. What impression did you have?

A. What it means to me is that he was not unalterably opposed, but on sum, adding up everything, he thought it would have been a mistake at that time to proceed with a crash program with all that entailed with this object that we didn't understand, when we had an awfully good program on hand in the fission field, which we did not wish to jeopardize. At least we did not feel it should be jeopardized. It turned out in the events that both could be done. Los Alamos just simply rose to the occasion and worked miracles, absolute miracles.

Mr. Marks. That is all.

e. Edward Teller.

Q. Dr. Teller, you know Dr. Oppenheimer well; do you not?

A. I have known Dr. Oppenheimer for a long time. I first got closely associated with him in the summer of 1942 in connection with atomic energy work. Later in Los Alamos and after Los Alamos I knew him. I met him frequently, but I was not particularly closely associated with him, and I did not discuss with him very frequently or in very great detail matters outside of business matters.

Q. To simplify the issues here, perhaps, let me ask you this question: Is your intention in anything that you are about to testify to, to suggest that Dr. Oppenheimer is disloyal to the United States?

A. I do not want to suggest anything of the kind. I know Oppenheimer as an intellectually most alert and very complicated person, and I think it would be presumptuous and wrong on my part if I would try in any way to analyze his motives. But I have always assumed, and I now assume that he is loyal to the United States. I believe this, and I shall believe it until I see very conclusive proof to the opposite.

Q. Now, a question which is the corollary of that. Do you or do you not believe that Dr. Oppenheimer is a security risk?

A. In a great number of cases I have seen Dr. Oppenheimer act—I understood that Dr. Oppenheimer acted—in a way which for me was exceedingly hard to understand. I thoroughly disagreed with him in numerous issues and his actions frankly appeared to me confused and complicated. To this extent I feel that I would like to see the vital interests of this country in hands which I understand better, and therefore trust more.

In this very limited sense I would like to express a feeling that I would feel personally more secure if public matters would rest in other hands.

---

36. Findings on the Case of J. Robert Oppenheimer: K. D. Nichols to the USAEC, June 12, 1954

United States Atomic Energy Commission
Washington 25, D.C.
June 12, 1954

Memorandum for: Mr. Strauss, Dr. Smyth, Mr. Murray,
Mr. Zuckert, Mr. Campbell

Subject: Dr. J. Robert Oppenheimer.

On December 23, 1953, Dr. J. Robert Oppenheimer was notified that his security clearance had been suspended, and informed of his right to a hearing under AEC procedures. By telegram dated January 29, 1954, Dr. Oppenheimer requested that he be afforded a hearing and on March 4,

This document appeared in *In the Matter of J. Robert Oppenheimer* (Washington, D.C.: Government Printing Office, 1954) pp. 43–48.

1954, after requesting and receiving three extensions of time, he submitted his answer to my letter of December 23, 1953.

Mr. Gordon Gray, Mr. Thomas A. Morgan and Dr. Ward V. Evans agreed to serve as members of the Personnel Security Board to hear Dr. Oppenheimer's case. The Board submitted its findings and recommendation to me on May 27, 1954. A majority of the Board recommended against reinstatement of clearance, Dr. Evans dissenting.

On May 28, 1954, I notified Dr. Oppenheimer of the recommendation of the Personnel Security Board and forwarded to him a copy of the Board's findings and recommendation. I informed Dr. Oppenheimer of his right to request review of his case by the Personnel Security Review Board and informed him that upon full consideration of the entire record in the case, including the recommendation of the Personnel Security Review Board in the event he requested review by that Board, I would submit to the Commission my recommendation as to whether or not his clearance should be reinstated. I also informed him that the final determination would be made by the Commission.

By letter of June 1, 1954, Dr. Oppenheimer waived his right to a review of his case by the Personnel Security Review Board and requested an immediate consideration of his case by the Commission.

FACTORS CONSIDERED

In making my findings and determination I have considered the question whether a security risk is involved in continued clearance of Dr. Oppenheimer. I have taken into account his contributions to the United States atomic energy program and in addition I have, in accordance with AEC procedures, considered the effect which denial of security clearance would have upon the program.

DR. OPPENHEIMER'S WORLD WAR II CONTRIBUTION

Dr. Oppenheimer has been intimately associated with the atomic energy program virtually from its inception. He participated in early weapons research and was selected as the wartime Director of the Los Alamos Laboratory. As district engineer of the wartime Manhattan Engineer District I was keenly aware of the contribution he made to the initial development of the atomic bomb. His leadership and direction of the Los Alamos weapons program were outstanding; his contributions leading to a successful atomic weapon have properly received worldwide acknowledgement and acclaim.

WORLD WAR II CLEARANCE 1943

As deputy district engineer of the Manhattan District, I was also aware of the circumstances, which have been brought out in the record, surrounding Dr. Oppenheimer's appointment as head of the Los Alamos Laboratory, and his subsequent clearance. He was selected in spite of the fact that he was considered a "calculated risk." He would not have been chosen had he not been considered virtually indispensable to the atomic bomb program. After he was chosen, as General Groves testified before the Board, Dr. Oppenheimer probably would not have been cleared had he not already been thoroughly steeped in knowledge of weapons research and had he not been considered absolutely essential.

Security officers opposed the clearance of Dr. Oppenheimer and it was not until July of 1943, after he had participated in the program for many months, that the decision to clear him was made by General Groves. I personally signed the directive advising the commanding officer at Los Alamos that there was no objection to Dr. Oppenheimer's employment.

The Manhattan District had one mandate—to build an atomic bomb as quickly as possible. Fears that Germany would build an atomic weapon first and possibly win the war thereby spurred the Manhattan District in what was felt to be a race against the German effort. Communist Russia was also fighting Germany at that time.

General Groves testified before the Board that he did not regret having made the decision to clear Dr. Oppenheimer in consideration of all of the circumstances which confronted him in 1943 but that under the present requirements of the Atomic Energy Act, as he interprets them, he would not clear Dr. Oppenheimer today.

BASIS FOR CLEARANCE UNDER ATOMIC ENERGY ACT
AND EXECUTIVE ORDER 10450

In this case as well as in all personnel security cases, the AEC in granting or reinstating a clearance must determine that the common defense and security will not be endangered. Under the Atomic Energy Act, such determination must be made on the basis of the character, associations, and loyalty of the individual concerned. Thus, a finding of loyalty in any given case does not suffice; substantial deficiency in any one of the three factors—character, associations, or loyalty—may prevent the determination that permitting such person to have access to restricted data will not endanger the common defense or security.

In addition, the criteria set up by Executive Order 10450 must be considered. This order requires that a program be established to insure that the retention in employment of any employee is clearly consistent with the interests of national security.

SECURITY FINDINGS

I have reviewed the entire record of the case, including the files, the transcript of the hearing, the findings and recommendation of the Personnel Security Board, and the briefs filed by Dr. Oppenheimer's attorneys on May 17, 1954, and June 7, 1954, and have reached the conclusion that to reinstate the security clearance of Dr. Oppenheimer would not be clearly consistent with the interests of national security and would endanger the common defense and security.

I concur with the findings and recommendation of the majority of the Personnel Security Board and submit them in support of this memorandum. In addition, I refer in particular to the following considerations:

1. Dr. Oppenheimer's Communist activities.—The record contains no direct evidence that Dr. Oppenheimer gave secrets to a foreign nation or that he is disloyal to the United States. However, the record does contain substantial evidence of Dr. Oppenheimer's association with Communists, Communist functionaries, and Communists who did engage in espionage. He was not a mere "parlor pink" or student of communism as a result of immaturity and intellectual curiosity, but was deeply and consciously involved with hardened and militant Communists at a time when he was a man of mature judgement.

His relations with these hardened Communists were such that they considered him to be one of their number. He admits that he was a fellow traveler, and that he made substantial cash contributions direct to the Communist Party over a period of 4 years ending in 1942. The record indicates that Dr. Oppenheimer was a Communist in every respect except for the fact that he did not carry a party card.

These facts raise serious questions as to Dr. Oppenheimer's eligibility for clearance reinstatement.

It is suggested that Dr. Oppenheimer has admitted many of the facts concerning his past association with Communists and the Communist Party. Whether this be true or not, it appears to me that Dr. Oppenheimer's admissions in too many cases have followed, rather than preceded, investigation which developed the facts. It appears that he is not inclined to disclose the facts spontaneously, but merely to confirm those already known. I find no great virtue in such a plea of guilt; certainly it does not cause me to dismiss Dr. Oppenheimer's past associations as matters of no consequence simply on the ground that he has admitted them.

2. The Chevalier incident.[1]—Dr. Oppenheimer's involvement in the Chevalier incident, and his subsequent conduct with respect to it, raise grave questions of security import.

If in 1943, as he now claims to have done, he knowingly and willfully made false statements to Colonel Pash, a Federal officer Dr. Oppenheimer violated what was then Section 80, Title 18, of the United States Code;[2] in

1. In 1943, while in charge of the Los Alamos Laboratory, Dr. Oppenheimer was interviewed by Colonel Pash, an officer of the U.S. Army Military Intelligence charged with the duty of protecting the atomic weapons project against spies. Dr. Oppenheimer told Colonel Pash, in detail, of an attempt on behalf of Soviet espionage to obtain from him information about the work on the atom bomb. Although he refused to disclose to Colonel Pash the name of the person making the approach, he subsequently identified him as Haakon Chevalier. Chevalier was a professional colleague and, at that time, a friend of Dr. Oppenheimer. The record includes Dr. Oppenheimer's testimony that his conversation with Chevalier about transmitting technical information to Soviet personages was brief and innocuous but that when questioned about it by Colonel Pash, he invented "a cock-and-bull story" involving the Russian Consulate, the use of microfilm, and approaches to three persons in the atom bomb project. The testimony (transcript pages 137 and 149) reads in part as follows:

Q. Now let us go back to your interview with Colonel Pash. Did you tell Pash the truth about this thing?
A. No.
Q. You lied to him?
A. Yes.
Q. Why did you go into such great circumstantial detail about this thing if you were telling a cock and bull story?
A. I fear that this whole thing is a piece of idiocy. I am afraid I can't explain why there was a consul, why there was microfilm, why there were three people on the project, why two of them were at Los Alamos. All of them seems wholly false to me.
Q. You will agree, would you not, sir, that if the story you told to Colonel Pash was true, it made things look very bad for Mr. Chevalier?
A. For anyone involved in it, yes, sir.
Q. Including you?
A. Right.
Q. Isn't it a fair statement today, Dr. Oppenheimer, that according to your testimony now you told not one lie to Colonel Pash, but a whole fabrication and tissue of lies?
A. Right.
Q. In great circumstantial detail, is that correct?
A. Right.

2. 18 U.S. Code, sec. 80, provides in pertinent part: "Whoever shall knowingly or willfully falsify or conceal or cover up by any trick, scheme, or device a material fact, or make or cause to be made any false or fraudulent statement or representations in any matter within the jurisdiction or agency of the United States shall be fined not more than $10,000 or imprisoned not more than 10 years, or both."

other words if his present story is true then he admits he committed a felony in 1943. On the other hand, as Dr. Oppenheimer admitted on cross-examination, if the story Dr. Oppenheimer told Colonel Pash was true, it not only showed that Chevalier was involved in a criminal espionage conspiracy, but also reflected seriously on Dr. Oppenheimer himself.

After reviewing both the 16-page transcript (as accepted by the Board) of the interview between Dr. Oppenheimer and Colonel Pash on August 26, 1943, and recent testimony before the Board, it is difficult to conclude that the detailed and circumstantial account given by Dr. Oppenheimer to Colonel Pash was false and that the story now told by Dr. Oppenheimer is an honest one. Dr. Oppenheimer's story in 1943 was most damaging to Chevalier. If Chevalier was Dr. Oppenheimer's friend and Dr. Oppenheimer, as he now says, believed Chevalier to be innocent and wanted to protect him, why then would he tell such a complicated false story to Colonel Pash? This story showed that Chevalier was not innocent, but on the contrary was deeply involved in an espionage conspiracy. By the same token, why would Dr. Oppenheimer tell a false story to Colonel Pash which showed that he himself was not blameless? Is it reasonable to believe a man will deliberately tell a lie that seriously reflects upon himself and his friend, when he knows that the truth will show them both to be innocent?

It is important to remember also that Dr. Oppenheimer did not give his present version of the story until 1946, shortly after he had learned from Chevalier what Chevalier himself had told the FBI about the incident in question. After learning of this from Chevalier, Dr. Oppenheimer changed his story to conform to that given to the FBI by Chevalier.

From all of these facts and circumstances, it is fair inference that Dr. Oppenheimer's story to Colonel Pash and other Manhattan District officials was substantially true and that his later statement on the subject to the FBI, and his recent testimony before the Personnel Security Board, were false.

Executive Order 10450, provides:

"Section 8 (a). The investigations conducted pursuant to this order shall be designed to develop information as to whether the employment or retention in employment in the Federal service of the person being investigated is clearly consistent with the interests of the national security. Such information shall relate, but shall not be limited, to the following:

(I) Depending on the relation of the Government employment to the national security:

(i) Any behavior, activities, or associations which tend to show that the individual is not reliable or trustworthy.

(ii) Any deliberate misrepresentations, falsifications, or omission of material facts.

(iii) Any criminal, infamous, dishonest, immoral, or notoriously disgraceful conduct, habitual use of intoxicants to excess, drug addiction, or sexual perversion."

In my opinion, Dr. Oppenheimer's behavior in connection with the Chevalier incident shows that he is not reliable or trustworthy; his own testimony shows that he was guilty of deliberate misrepresentations and falsifications either in his interview with Colonel Pash or in his testimony before the Board; and such misrepresentations and falsifications constituted criminal, dishonest conduct.

Further, the significance of the Chevalier incident combined with Dr. Oppenheimer's conflicting testimony from 1943 to 1954 in regard to it were not, of course, available in whole to General Groves in 1943, nor was the complete record of the Chevalier incident considered by the Atomic Energy Commission in 1947. Consideration of the complete record plus a cross-examination of Dr. Oppenheimer under oath were not accomplished by anyone prior to the Personnel Security Board hearing in 1954.

3. Dr. Oppenheimer's veracity.—A review of the record reveals other instances which raise a question as to the credibility of Dr. Oppenheimer in his appearance before the Personnel Security Board and as to his character and veracity in general.

(a) The record suggests a lack of frankness on the part of Dr. Oppenheimer in his interview with the FBI. It appears that during this hearing he recollected details concerning Communist meetings in the San Francisco area which he did not report in previous interviews with the FBI.

(b) Dr. Oppenheimer told the FBI in 1950 that he did not know that Joseph Weinberg was a Communist until it became a matter of public knowledge. When confronted with the transcript of his interview with Colonel Lansdale on September 12, 1943, he admitted that he had learned prior to that date that Weinberg was a Communist.

(c) It is clear from the record that Dr. Oppenheimer was a great deal more active in urging the deferment of Rossi Lomanitz and his retention on the atom bomb project than he said he was in his answer to my letter of December 23, 1953. Furthermore, Dr. Oppenheimer testified that if he had known that Lomanitz was a Communist he would not have written the letter to Colonel Lansdale of the Manhattan District on October 19, 1943, supporting Lomanitz' services for the project. However, the record reflects that Dr. Oppenheimer told Colonel Lansdale of the Manhattan District on September 12, 1943, that he had learned that Lomanitz was a Communist.

(d) Dr. Oppenheimer admitted in his testimony before the Board that in 1949 he wrote a letter to a newspaper which might have misled the public concerning his testimony before the House Un-American Activities Committee on Dr. Bernard Peters. He testified that an earlier article in the newspaper which summarized his testimony was accurate, yet the effect of his published letter was to repudiate the earlier article.

(e) Dr. Oppenheimer in his answer to my letter of December 23, 1953, and in his testimony before the Board with respect to the H-bomb program undertook to give the impression that in 1949 he and the GAC merely opposed a so-called "crash" program. It is quite clear from the record, however, that the position of the majority of the GAC, including Dr. Oppenheimer, was that a thermonuclear weapon should never be produced, and that the United States should make an unqualified commitment to this effect. In discussing the building of neutron-producing reactors, a majority of the GAC, including Dr. Oppenheimer expressed the opinion that "the super program itself should not be undertaken and that the Commission and its contractors understand that the construction of neutron-producing reactors is not intended as a step in the super program." The testimony of Dr. Oppenheimer viewed in light of the actual record certainly furnished adequate basis for the majority of the Board not believing that Dr. Oppenheimer was entirely candid with them on this point.

(f) Dr. Oppenheimer testified before the Board that the GAC was unanimous in its basic position on the H-bomb. He specifically said that Dr. Seaborg had not expressed his views and that there was no communication with him. It should be noted that the statement that "there was no communication with him" was volunteered by Dr. Oppenheimer in his testimony on cross-examination before the Board. However, Dr. Oppenheimer received a letter from Dr. Seaborg, expressing his views, prior to the October 29, 1949, GAC meeting.

Dr. Oppenheimer's continued associations after World War II. —Dr. Oppenheimer has continued associations which raise a serious question as to his eligibility for clearance. He has associated with Chevalier on a rather intimate basis as recently as December 1953, and at that time lent his name to Chevalier's dealings with the United States Embassy in Paris on a problem which, according to Dr. Oppenheimer, involved Chevalier's clearance. Since the end of World War II he has been in touch with Bernard Peters, Rossi Lomanitz, and David Bohm under circumstances which, to say the least, are disturbing.

Obstruction and disregard of security. —Dr. Oppenheimer's actions have shown a consistent disregard of a reasonable security system. In addition to the Chevalier incident, he has refused to answer questions put to him by security officers concerning his relationships and knowledge of particular individuals whom he knew to be Communists; and he has repeatedly exercised an arrogance of his own judgment with respect to the loyalty and reliability of his associates and his own conduct which is wholly inconsistent with the obligations necessarily imposed by an adequate security system on those who occupy high positions of trust and responsibility in the Government.

FINDING OF SECURITY RISK IS NOT BASED ON DR. OPPENHEIMER'S OPINIONS

Upon the foregoing considerations relating to the character and associations of Dr. Oppenheimer, I find that he is a security risk. In making this finding I wish to comment on the item of derogatory information contained in my letter of December 23, 1953 which relates to the hydrogen bomb and, in particular, which alleged that:

"It was further reported that even after it was determined, as a matter of national policy, to proceed with development of a hydrogen bomb, you continued to oppose the project and declined to cooperate fully in the project. It was further reported you departed from your proper role as an adviser to the Commission by causing the distribution, separately and in private, to top personnel at Los Alamos of the majority and minority reports of the General Advisory Committee on development of the hydrogen bomb for the purpose of trying to turn such top personnel against the development of the hydrogen bomb. It was further reported that you were instrumental in persuading other outstanding scientists not to work on the hydrogen bomb project, and that the opposition to the hydrogen bomb, of which you are the most experienced, most powerful and most effective member, has definitely slowed down its development."

It should be emphasized that at no time has there been any intention on my part or the Board's to draw in question any honest opinion expressed by Dr. Oppenheimer. Technical opinions have no security implications unless they are reflections of sinister motives. However, in view of Dr. Oppenheimer's record coupled with the preceding allegation concerning him, it was necessary to submit this matter for the consideration of the Personnel Security Board in order that the good faith of his technical opinions might be determined. The Board found that, following the President's decision, Dr. Oppenheimer did not show the enthusiastic support for the program which might have been expected of the chief atomic adviser to the Government under the circumstances; that, had he given his enthusiastic support to the program, a concerted effort would have been initiated

at an earlier date, and that whatever the motivation, the security interests
of the United States were affected. In reviewing the record I find that the
evidence establishes no sinister motives on the part of Dr. Oppenheimer
in his attitude on the hydrogen bomb, either before or after the President's
decision. I have considered the testimony and the record on this subject
only as evidence bearing upon Dr. Oppenheimer's veracity. In this context
I find that such evidence is disturbing.

DR. OPPENHEIMER'S VALUE TO ATOMIC ENERGY
OR RELATED PROGRAMS

In addition to determining whether or not Dr. Oppenheimer is a security risk, the General Manager should determine the effect which denial of
security clearance would have upon the atomic energy or related programs.
In regard to Dr. Oppenheimer's net worth to atomic energy projects, I
believe, first, that through World War II he was of tremendous value and
absolutely essential. Secondly, I believe that since World War II his value
to the Atomic Energy Commission as a scientist or as a consultant has
declined because of the rise in competence and skill of other scientists and
because of his loss of scientific objectivity probably resulting from the diversion of his efforts to political fields and matters not purely scientific in
nature. Further, it should be pointed out that in the past 2 years since he
has ceased to be a member of the General Advisory Committee, his services have been utilized by the Atomic Energy Commission on the following occasions only:

October 16 and 17, 1952.
September 1 and 2, 1953.
September 21 and 22, 1953.

I doubt that the Atomic Energy Commission, even if the question of his
security clearance had not arisen, would have utilized his services to a
markedly greater extent during the next few years. I find, however, that
another agency, the Science Advisory Committee, Office of Defense Mobilization, has stated in a letter dated June 4, 1954, signed by Dr. L. A.
DuBridge, Chairman, and addressed to Chairman Strauss, that:
    "It is, therefore, of great importance to us that Dr. Oppenheimer's 'Q'
clearance be restored. This is especially true since our Committee is planning to undertake during the coming months an intensive study of impor-

tant items related to national security on which Dr. Oppenheimer's knowledge and counsel will be of very critical importance."

Dr. DuBridge further stated that:

"His value is, it seems to me, so enormous as to completely overbalance and override the relatively trivial risks[3] which the Personnel Security Board reports. In other words, the net benefits to national security will be far greater if Dr. Oppenheimer's clearance is restored than if it is terminated. Even though he served the Government in no other capacity than as a member of the Science Advisory Committee of the Office of Defense Mobilization, the above statements will, I am confident, be true beyond question."

Other Government agencies may also desire to use Dr. Oppenheimer's services if he were to be cleared. In addition, contractors and study groups involved in atomic power activities undoubtedly will from time to time, as one or two have already indicated, desire to clear Dr. Oppenheimer for consulting work.

Dr. Oppenheimer could of course make contributions in all these fields, but he is far from being indispensable.

CONCLUSION

I have conscientiously weighed the record of Dr. Oppenheimer's whole life, his past contributions, and his potential future contributions to the Nation against the security risk that is involved in his continued clearance. In addition, I have given consideration to the nature of the cold war in which we are engaged with communism and Communist Russia and the horrible prospects of hydrogen bomb warfare if all-out war should be forced upon us. From these things a need results to eliminate from classified work any individuals who might endanger the common defense or security or whose retention is not clearly consistent with the interests of national security.

Dr. Oppenheimer's clearance should not be reinstated.

K. D. Nichols

General Manager.

---

3. It should be noted that Dr. DuBridge to my knowledge has never had access to the complete file or the transcript of the hearing.

37. Arthur Holly Compton to Gordon Gray, April 21, 1954, "Evidence Regarding Loyalty of Robert Oppenheimer"

Park Hotel
Istanbul, Turkey

April 21, 1954

Dear President Gray:

I was in Pakistan when the news broke regarding the suspension of J. Robert Oppenheimer from the Advisory Board of the A.E.C., and it was not until yesterday that I received sufficiently full information about the case to make a statement that I knew would be relevant. Because of my close association with Oppenheimer, I believe that some facts from my first-hand knowledge will be useful to your committee in judging his loyalty. These facts I am stating here with the same responsibility for their veracity as I would accept in the case of sworn testimony before a court of law.

1. I was responsible for appointing Oppenheimer to the task of organizing whatever was necessary for designing the atomic bomb. This was about the end of April, 1942. (The dates given here are all approximate. I have here no notes regarding these matters, and all of my statements are from memory.) It was a month or two later that responsibility for the atomic development was assigned to the U.S. Engineers (Army), and it was in September, 1942, that I recommended to General Groves that Oppenheimer be continued at this task, and that the responsibility be enlarged to include the construction of the bomb. This recommendation was in accord with that of others, in particular that of Dr. James Conant, and was accepted.

At the time of Oppenheimer's appointment the atomic development was, as you know, in civilian hands. By Dr. Conant, acting for the OSRD, I was given complete responsibility for such appointments, reporting to him, however, regarding actions taken. For my guidance, within our "Metallurgical Project," a personnel division had been established which investigated the loyalty background of employees. This investigation (and our other security measures) was carried on with the help of the Federal Bu-

This document appeared in the Arthur Holly Compton Papers, Washington University in St. Louis, Archives, Olin Library.

reau of Investigation. But we kept within our own hands the responsibility for the decisions in each case.

It was recognized that the task to which we were appointing Oppenheimer was not only of essential importance to the success of the total project, but also one in which a unique degree of discretion was required. It was my judgment that Oppenheimer's qualifications fitted him for such a post better than any other person who could be made available. This was after a diligent search for about a month which included consultation with many top-level scientists.

That Oppenheimer had had contacts with Communists was well known to me. These contacts, as I knew them, were essentially as described in his letter as published in *The New York Times* of April 13. It was my impression that, as one eager to find a solution to world problems, he had investigated Communism first-hand to see what it had to offer.

The much more significant fact was that he had already become disillusioned with Communism by what he had found. In a conversation with him about another matter, in the spring of 1940, I believe, he had given me his reasons for not associating himself with a professional organization that had some Communist ties. Later, in 1941, I believe, he had told me of his efforts to persuade his brother Frank to dissociate himself from Communist groups. As I recall it, it was during the early stages of our conversations about the atomic program, before I had approached him about accepting the responsibility for the design of the bomb, that Oppenheimer told me he was breaking completely every association with any organization that might be suspected of Communism, in order that he might be of maximum usefulness on war projects.

It was my judgment that a person who had known Communism and had found its faults was more to be relied upon than one who was innocent of such connections. This was the more important because it was evident that his task would necessarily draw in many men of foreign background, who were among those most competent in theoretical physics. Without the use of such men his task could hardly have been accomplished. A man with Oppenheimer's experience both in foreign countries and, in a limited way, with Communism, was in a most favorable position to recognize those whose loyalties might be directed elsewhere than to the United States and the free world.

The chief positive reason for selecting Oppenheimer for this post was that after working vigorously on rallying the help of theoretical physicists to the design of the bomb, Oppenheimer had shown the most eagerness and initiative, he was one of the very few American-born men who had the professional competence, and he had demonstrated a certain firmness of character. While his administrative competence remained to be demonstrated, it looked to me promising.

Looking back on this selection, I do not believe it would have been possible to find anywhere a man better suited to carry through this unique job in the nation's interest.

2. After the war was over I discussed with Oppenheimer several times the question of proceeding with the development of the H-bomb. Its possibility Oppenheimer had called to my attention in August, 1942. This possibility was kept under the utmost secrecy; but before the war was over its suggestion had arisen spontaneously from so many sources that it was evident that it would arise in any group working seriously on nuclear explosions. In particular, it was evident that the basic principle of the fusion bomb would be known in Russia.

Immediately after the war our great consideration was to reach a reliable agreement with Russia that would rule out atomic weapons but permit the development of atomic power. During this period there was no immediate occasion to work toward atomic fusion. The situation changed when Russia showed such reluctance that no atomic weapon agreement seemed feasible. We then recognized that Russia was working vigorously on her own atomic development. Presumably the H-bomb would eventually become part of her program.

I recall a conversation with Oppenheimer in 1947 or 48 in which I was advocating the initiation of a program of active research and development toward the H-bomb. My point was that this development, if it was physically possible, was sure to come before many years, and it was important that the availability of this "super" weapon should first be in our hands.

I found Oppenheimer reluctant. His chief reluctance was, I believe, on purely moral grounds. No nation should bring into being a power that would (or could) be so destructive of human lives. Even if another nation should do so, our morality should be higher than this. We should accept the military disadvantage in the interest of standing for a proper moral principle.

He had other reasons—the development of fear and antagonism among other nations, the substantial possibility that the effort to create an atomic explosion would fail, questions regarding the H-bomb's military value. He hoped that no urgent need for its development would arise.

In this and other conversations Oppenheimer brought up precisely those questions that needed to be considered. His thinking seemed to be aimed solely toward finding what was in the best interests of the United States. He took for granted, as did I, that the United States' interests are those of humanity. There was no shadow of a suspicion that his arguments were subtly working toward Russia's advantage. I am confident that no such thought was in his mind.

With the explosion of Russia's first atomic bomb in 1949 the situation was sharply changed. I do not recall any first-hand discussions with Oppenheimer on this matter after that date.

3. Having known Robert Oppenheimer since his days in Gottingen in 1927, having worked with him closely during the war years, and having kept in touch with him occasionally since, it is my judgment that he is completely loyal to the interests of the United States, and that any activity in the interest of a foreign power at the expense of the United States would be thoroughly repugnant to him. It is my judgment further that he is, and has been since 1941, just as thoroughly opposed to Communism.

Yours sincerely,
Arthur H. Compton

AHC:dbs
(Permanent address,
Washington University,
St. Louis, MO. U.S.A.)

VI

Nuclear Testing and the Test Ban

*I*n July 1946, just two weeks after Bernard Baruch had presented his plan for international control of atomic weapons, a B-29 dropped an atomic bomb on a fleet of captured Japanese naval vessels in the lagoon at Bikini Atoll in the Pacific Ocean. Several weeks later another weapon was exploded under the water. Operation Crossroads, as the two nuclear tests were named, were the first nuclear explosions in peacetime. Before the signing of the Limited Test Ban Treaty in 1963, the United States would announce 328 additional nuclear tests.

Coinciding as it did with the presentation of the Baruch Plan, Operation Crossroads would spark charges against the United States of conducting atomic diplomacy. The government, however, had discussed test plans since the end of the war. Like other parts of the armed services, the program in atomic weapons was also being demobilized, and the military wanted to test the effects of atomic bombs on ships and other equipment before the scientists returned to their universities.

Shortly after Crossroads, the civilian Atomic Energy Commission assumed responsibility for the nuclear program. The Commission learned that the Hiroshima-type bomb was too inefficient for the short supply of fissionable material. Moreover, the Nagasaki-type weapon was more a laboratory than a production model. Consequently, the Commission focused on rebuilding its laboratories, its plutonium production facilities, and its testing program. By the spring of 1948 the Commission conducted its first atomic tests, codenamed Operation Sandstone, on Eniwetok, 180 miles west of Bikini. The primary purpose of Sandstone was to improve the efficiency of weapon design so that scarce fissionable material would be better utilized and the bomb size and weight could be reduced.

Introduction

To most Americans the rush of world events justified the Atomic Energy Commission's nuclear testing program. The Soviet rejection of the Baruch Plan, Soviet satellite governments in control in eastern Europe, the Greek Civil War, the Communist coup in Czechoslovakia, and the Berlin Blockade in 1948 heralded the realities of the Cold War. Within the next year Americans learned that Klaus Fuchs, a British scientist who had worked on the bomb at Los Alamos, had sold atomic secrets to the Russians. In addition, Mao Tse-tung's Communist armies had captured the Chinese mainland, and the Soviet Union had detonated its first atomic bomb. These events, especially the news of the Soviet atomic bomb, quickened a debate within the Atomic Energy Commission and between high government officials over the development of a thermonuclear weapon.

The North Korean invasion of South Korea in June 1950 appeared to justify President Truman's decision to proceed with a "super" bomb. The "police action" might become a full-scale war in which nuclear weapons would be the deciding factor.

The logistics of testing new weapons designs in the Pacific placed limitations on the efficiencies and costs of the testing program. Scientists at Los Alamos argued for a testing site within the United States. The threat that the Eniwetok site might be cut off in a war brought a new urgency to open a continental test location. By late 1950 the government decided on a large area in Nevada, north of Las Vegas. Although the government located the site in a remote area, the Atomic Energy Commission now confronted the hazard of radioactive fallout within the United States. Because neither Greenhouse nor Sandstone had involved radiological hazards to nearby populations, precise information concerning the effect of radiation on people was less important than weapons data. Tests on the American continent, however, could pose real dangers to the general public.

Continental testing began in January 1951 with Operation Ranger, a series of four shots, followed by Operation Buster-Jangle in the fall. The weapons, mostly dropped from aircraft, did not produce significant amounts of fallout off the test site. However, fallout did occur in Rochester, New York, exposing film at the Eastman Kodak Company and causing the commission to establish a long-range radiation monitoring system throughout the United States.

The drive to produce a thermonuclear, or hydrogen, bomb led to further experimentation at the Nevada Test Site and in the Pacific. At Operation Ivy at Eniwetok in November 1952, a prototype thermonuclear device named Mike yielded an explosive power of over ten megatons (ten million tons of TNT) and proved that a hydrogen bomb was possible. But Mike weighed over eight tons and was not a deliverable weapon. That was the next task.

To obtain precise measurements for designing and testing of hydrogen weapons in the Pacific (the Commission prohibited thermonuclear testing in Nevada), the continental tests carried out during Upshot-Knothole in the spring of 1953 were fired from 300-foot towers. As a result, for the first time the atomic fireball swept up great amounts of debris from the desert floor and deposited significant levels of radioactive fallout on Utah and Nevada. According to some estimates, 80 percent of offsite fallout from continental testing came from the Upshot-Knothole series.

The high levels of fallout caused the Atomic Energy Commission to reexamine continental testing and recommend tighter restrictions on test procedures, especially weather conditions, which largely determined where and how much fallout would occur. But two months after Upshot-Knothole concluded, the Soviet Union detonated its first thermonuclear device. The race for weapons of greater destruction intensified.

By 1954 the Atomic Energy Commission had established a second weapons laboratory in Livermore, California, and had made plans for a large thermonuclear test on Bikini, which had not been used since Crossroads. The Castle series, designed by the Los Alamos laboratory, would test a weapon so powerful that most civilian and military personnel were evacuated to ships thirty miles from ground zero.

The preparations were well taken but inadequate. Shot Bravo yielded fifteen megatons and spread radioactive fallout far beyond the thirty-mile perimeter. Ships received heavy radiation and sailed out to fifty miles. Later, when part of the task force returned to Bikini, it could not enter the area due to high radiation levels. Bravo also exposed natives on islands a hundred miles away. In addition, the fallout hit a Japanese fishing vessel, the Lucky Dragon, and brought worldwide attention to the dangers of thermonuclear weapons.

That American troops were exposed to fallout at Bravo was not a new situation. Since 1952 the Department of Defense had placed Army troops at the Nevada Test Site to test their reactions during a nuclear attack. Seven thousand soldiers had participated at Operation Tumbler-Snapper (1952) and more than 17,000 at Upshot-Knothole (1953). Psychiatrists charted troop reactions to the blast and their ability to perform simple tasks and brief maneuvers following the explosion. In addition, troops participated at Operation Teapot in 1955 and Operation Plumbbob in 1957.

Even as nuclear weapons development and testing proceeded, President Eisenhower in 1954 asked his aides to consider a moratorium on nuclear testing. Opposition from the Departments of Defense and State and the Atomic Energy Commission doomed this initial attempt. But Eisenhower appointed Harold Stassen as a special adviser on disarmament, and Stassen, with *Henry Cabot Lodge, the ambassador to the United Nations, continued to explore Eisenhower's wishes for a test ban and disarmament.*

In 1956 Adlai Stevenson, the Democratic candidate for president, spoke vigorously for a test ban agreement, but Eisenhower, angry that disarmament had become a campaign issue, refused to have a test ban without a general disarmament agreement. Americans, troubled by instability abroad—the Soviet invasion of Hungary and the British-French-Israeli invasion of Egypt—agreed with Eisenhower, who easily won reelection.

Still, Eisenhower hoped to end nuclear testing and announced in the summer of 1958 that the United States would voluntarily suspend weapons testing. There was a rush of testing in Operation Hardtack, *first in the Pacific and then in Nevada, before the moratorium went into effect on October 31, 1958. Since Alamogordo, the United States had detonated 151 atomic devices.*

On August 31, 1961, Chairman Nikita Khrushchev announced that the Soviet Union would begin nuclear testing. Two weeks later the United States followed suit. Most tests in Nevada took place underground, except for cratering experiments done for the Plowshare program to analyze the peaceful uses of nuclear explosions (such as earth moving or digging a canal).

Two years later, in August 1963, President John F. Kennedy joined Great Britain and the Soviet Union in signing the Limited Test Ban Treaty prohibiting nuclear testing in the atmosphere, underwater, or in space. Since then, all American nuclear tests have been conducted underground.

On testing and events leading up to the 1963 test ban, see Robert A. Divine, Blowing on the Wind: The Nuclear Test Ban Debate, 1954–1960 *(New York: Oxford University Press, 1978).*

38. Announced Nuclear Tests, 1945–52

Series	Date (GCT)	Location	Type of Burst	Yield (Kilotons)	Remarks
		UNITED STATES TESTS			
Trinity	July 16, 1945	Alamogordo, N.M.	Tower	19	First nuclear detonation
World War II					
Little Boy	Aug. 5, 1945	Hiroshima, Japan	Air	12.5	First combat use

Continued

Series	Date (GCT)	Location	Type of Burst	Yield (Kilotons)	Remarks
		UNITED STATES TESTS			
Fat Man	Aug. 9, 1945	Nagasaki, Japan	Air	22	Second combat use
Crossroads					Test of nuclear
Able	June 30, 1946	Bikini Atoll	Air	Classified	weapon effects
Baker	July 24, 1946	Bikini Atoll	Underwater	Classified	on naval vessels
Sandstone					First Commis-
X-ray	Apr. 14, 1948	Eniwetok Atoll	Tower	37	sion test series
York	Apr. 30, 1948	Eniwetok Atoll	Tower	49	
Zebra	May 14, 1948	Eniwetok Atoll	Tower	18	
Ranger					First test held
Able	Jan. 27, 1951	Nevada test site	Air	1	at Nevada test
Baker	Jan. 28, 1951	Nevada test site	Air	8	site
Easy	Feb. 1, 1951	Nevada test site	Air	1	
Baker-2	Feb. 2, 1951	Nevada test site	Air	8	
Fox	Feb. 6, 1951	Nevada test site	Air	22	
Greenhouse					Demonstrated
Dog	Apr. 7, 1951	Eniwetok Atoll	Tower	Classified	thermonuclear
Easy	Apr. 20, 1951	Eniwetok Atoll	Tower	47	principles
George	May 8, 1951	Eniwetok Atoll	Tower	Classified	
Item	May 24, 1951	Eniwetok Atoll	Tower	Classified	
Buster-Jangle					Tests of
Able	Oct. 22, 1951	Nevada test site	Tower	0.1	improved
Baker	Oct. 28, 1951	Nevada test site	Air	3.5	device designs
Charlie	Oct. 30, 1951	Nevada test site	Air	14	
Dog	Nov. 1, 1951	Nevada test site	Air	21	
Easy	Nov. 5, 1951	Nevada test site	Air	31	
Sugar	Nov. 19, 1951	Nevada test site	Surface	1.2	
Uncle	Nov. 29, 1951	Nevada test site	Underground	1.2	
Tumbler-Snapper					Tests of improved
Able	Apr. 1, 1952	Nevada test site	Air	1	device designs
Baker	Apr. 15, 1952	Nevada test site	Air	1	

Reported Nuclear Test Explosions, 1945–68

Continued

Series	Date (GCT)	Location	Type of Burst	Yield (Kilotons)	Remarks
		UNITED STATES TESTS			
Charlie	Apr. 22, 1952	Nevada test site	Air	31	
Dog	May 1, 1952	Nevada test site	Air	19	
Easy	May 7, 1952	Nevada test site	Tower	12	
Fox	May 25, 1952	Nevada test site	Tower	11	
George	June 1, 1952	Nevada test site	Tower	15	
How	June 5, 1952	Nevada test site	Tower	14	
Ivy					First thermonu-
Mike	Oct. 31, 1952	Eniwetok Atoll	Surface	10,400	clear detonation
King	Nov. 15, 1952	Eniwetok Atoll	Air	Classified	
		FOREIGN TESTS			
United Kingdom					
Hurricane	Oct. 3, 1952	Monte bello Islands, Australia	Ship		Kiloton range
Soviet Union					
Test No. 1	Aug. 29, 1949	U.S.S.R.	Air		
Test No. 2	Oct. 3, 1951	U.S.S.R.	Air		
Test No. 3	Oct. 22, 1951	U.S.S.R.	Air		

This table appeared in Richard G. Hewlett and Francis Duncan, *Atomic Shield: A History of the USAEC*, vol. 2 (Washington, D.C.: Government Printing Office, 1972), pp. 672–73.

---

39. Reported Nuclear Test Explosions, 1945–68

	USA				USSR[d]						Total, all nations[e]
	AEC[i]										
	Test de- tection	Plow- share	Safety[b]	Weapon	Total	FOA[c]	AEC[i]	FOA[c]	UK	France	China	
Pre-PTBT[a]												
1945				1	1	1						1
1946				2	2	2						2
1947				–	–	–						–
1948				3	3	3						3

continued

USA / AEC[i] spans Test detection–Total; USSR[d] spans FOA[c], AEC[i].

	Test detection	Plowshare	Safety[b]	Weapon	Total	FOA[c]	AEC[i]	FOA[c]	UK	France	China	Total, all nations
1949				–	–	–	1	1				1
1950				–	–	–	–	–				–
1951				16	16	16	2	2				18
1952				10	10	10	–	–	1			11
1953				11	11	11	2	2	2			15
1954				6	6	6	1	2	–			8
1955				15	15	15	4	4	–			19
1956			1	13	14	14	7	8	6			28
1957			4	24	28	28	13	13	7			48
1958			14	52	66	66	25	27	5			98
Pre-1959								30[f]				30
1959				–	–	–	–	–				–
1960				–	–	–	–	–		3		3
1961		1		8	9	9	31	32	–	1		42
1962		3		86	89	88	40	42	2	2		135
1963 (pre-10 Oct.)		1		16	17	17	–	–	–	1		18
15 Sept. 1961– 20 Aug. 1963							23	23[g]				23
Total pre-PTBT		5	19	263	310	309	126	163	23	7		503
Annual rate of testing, pre-PTBT: 1951–1963					24.4		12.8					39.6
Post-PTBT												
1963 (post-10 Oct.)	1	1		7	9	8	–	–	–	1		10
1964	1	6		21	29	28	3	6	1	–	1	37
1965	1	1		24	27	27	4	9	1	2	1	40
1966	1	4		35	40	40	7	12	–	5	3	60
1967	–	3		25	28	28	4	13	–	3	2	46
1968	1	8[h]		28	37	37	6	9	–	5	1	52
Total, post-PTBT	5	23		140	170	168	24	49	2	16	8	245

continued

	USA					USSR[d]						Total, all
	AEC[i]											
	Test de-tection	Plow-share	Safety[b]	Weapon	Total	FOA[c]	AEC[i]	FOA[c]	UK	France	China	nations[e]
Annual rate of testing, post-PTBT: 1963–1968					32.0			9.2				46.2
Total, all tests	5	28	19	403	480	477	150	212	25	23	8	748

[a]Partial Test Ban Treaty (*10 October 1963*). [b]These are experiments to determine the safety of nuclear weapons in case of accident. [c]Swedish Research Institute for National Defence (Försvarets Forskning-sanstalt). [d]No official information is available for the Soviet Union. [e]When two sources give different figures, the higher of the two is taken. [f]These tests are reported by FOA as additional tests which took place at unspecified dates before 1959. [g]These tests are reported by the AEC as having taken place between 15 September 1961 and 20 August 1963. [h]Including 5 devices separately used in the same test (Project Buggy), counted here as 5. [i]Atomic Energy Commission.

This table appeared in *Stockholm Peace Research Institute Yearbook of World Armaments and Disarmament, 1968/1969* (New York: Humanities Press, 1969), p. 242.

---

40. "Civil Defense Faces New Perils," Ralph E. Lapp, 1954

Dr. Lap assesses the radioactive hazard of "fall-out" and analyzes its impact upon civil defense. His data are gathered from Japanese sources, from independent calculations, and from the unclassified scientific literature. He urges that the federal government release classified data on fall-out to provide guidance to civil defense organizations.

On March 1, 1954 chalk-white dust fell on twenty-three Japanese fishermen 72 miles from Bikini. It took three hours for the pulverized coral to start falling-out from the air and coat the *Lucky Dragon* tuna ship with a mantle of radioactive debris. However, it was not until late this summer that the Federal Civil Defense Administration felt the impact of "fall-out."

This essay first appeared in *Bulletin of the Atomic Scientists* 10 (November 1954): 349–51.

Q-cleared officials in FCDA were briefed on the nature of the fall-out. They were shown colored charts with neat elliptical contours describing the range of lethality of the residual radioactivity from superweapon explosions. Up to this point FCDA had worked and thought mostly in terms of circles—the symmetric patterns of primary damage from superbombs which the *Bulletin* published last month. Now superimposed upon the great circles of H-bomb blast and heat, there were zeppelin-shaped ellipses which stretched far beyond the circles of primary damage.

These ellipses stunned civil defense planners. By a major shift of policy they had replaced previous "duck-and-cover" or "stay put" planning by a policy of pre-attack evacuation which dispersed metropolitan populations beyond the inner circles of near-total and heavy blast damage. Gov. Val Peterson, as head of FCDA, made the policy switch once he took a good hard look at the blast and heat effects of our MIKE explosion of November 1, 1952. What Gov. Peterson realized very clearly was that MIKE was not the last word in superweapon development. Indeed a weapon twice its power was tested in the March-April CASTLE series of tests in the Pacific in 1954. Moreover, this weapon was a bomb, not a "device"—meaning that the United States now possesses a droppable bomb in the range of 20 megatons.

Faced with the prospect of corresponding weapons in the Soviet arsenal, Gov. Peterson recognized that too much of America's metropolitan population resided inside the 14 mile radius of the superbomb's punch. It was sheer suicide, he reasoned, to put 35 million Americans in the sitting-duck category. This is the simple background for the policy of evacuation which is now being implemented.

It was at this point in the evolution of C. D. policy that "fall-out" descended upon civil defense planners.

*   *   *

Now radioactivity was not a new wrinkle to the planners. The Baker shot of a 20 kiloton bomb in the Bikini lagoon in July of 1946 saturated the mushroom cloud with awesome quantities of radioactivity. Millions of tons of salt water were erupted into the air and a misty radioactive steam surged across the lagoon surface. An egg-shaped ellipse about 3 miles in length constituted the lethal area of this radioactivity. This area was not comparable to that of a metropolis, so as time went by radioactivity as a menace shrunk to less formidable proportions in civil defense planning.

The dimensions of the superbomb "fall-out" greatly exceeded those for a 1946 A-bomb explosion. Unfortunately, the exact or even rough forecasting of these dimensions is subject to a number of uncertainties. These can be appreciated by describing the fall-out from a superbomb detonation. When the bomb explodes there is a flash of penetrating radiation consisting of neutrons and gamma rays. Fortunately, the area subjected to lethal bombardment by this primary flash is small compared with that affected by the heat and blast of the explosion. Although altitude-dependent, the area is much less than that of total destruction from the blast wave.

Once the bomb explodes and the heat-blast waves have run their course, we have to consider the fate of the bomb cloud. If the bomb is burst high in the air so that there is no significant cratering effect the bomb cloud will contain only such surface debris as is sucked up into the ascending column. In this case, the cloud radioactivity consists predominantly of split uranium or plutonium atoms (technically called fission products) which are intensely radioactive. Two factors take their toll in reducing the menace of this activity. One—the high velocity upper air winds disperse the fine (usually invisible) particles in the bomb cloud so that they are dispersed over a very large area before they finally settle out and come to earth. Two—radioactive decay sharply reduces the activity on the following time schedule:

Time (after burst)	Radioactivity (arbitrary units)
1 minute	1,000,000
1 hour	7,300
1 day	162
1 week	16
1 month	3

Assuming that there were no mass deposits of radioactive debris in the early history of the bomb cloud, then we would have to relegate the problem of cloud activity to the controversial area of global atmospheric contamination and human genetics.

Superbombs which burst close to the surface present quite another problem. In this case, the 3¼ mile wide fire-ball of a 10-megaton bomb introduces a radically new factor into the fall-out equation. Much of the substratum below the exploding bomb is dislodged and volatilized into particles impregnated with radioactivity. In addition, some of the elements in the substratum may become radioactive by the primary penetrating radiation from the bomb. Sodium in sea water, for example, is easily activated (made radioactive) and can become a hazard. The pulverized substratum is funneled upward much in the manner of a cyclone. In this way, the coral and sand of a low-lying Bikini atoll island were sucked up into the bomb cloud.

As is well known the cauliflower cloud mushrooms upward to a height of over 10 miles in about 12 minutes. The characteristic 40 to 60 knot winds of the stratosphere then distort the mushroom in a downwind direction. At this point the difference between a high air burst and a low one becomes significant. The subsurface particles tend to fall out from the bomb cloud early in its history because of their massiveness. Fall-out of up to 50 per cent of the radioactive debris occurs in the first 24 hours with the maximum activity being deposited in the 1 to 3 hour period.

Estimation of the area of serious-to-lethal fall-out involves many unknowns. First, there is the power of the bomb. Second, the architecture of the weapon, involving principally the contribution of fission to the energy release in the bomb. Third, there is the height of the bomb burst. Fourth, one is confronted with the nature of the substratum—both as a carrier of the radioactivity and also as a producer of radioactivity. (In general, it would seem that induced activities cannot compete with the radioactivity produced in the bomb itself. However, a bomb burst close to sea water would produce vast quantities of radiosodium and radiochlorine. For example, a bomb burst in the Pacific Ocean off the coast of Los Angeles would probably cloak the city in a lethal fog even though no blast shook the city.) Finally, the meteorology at the time of the burst would determine the location and extent of the fall-out.

It is like seeking the Holy Grail to quest for hard data on the fall-out as applied to an American city. The unknowns enumerated above make prediction of the fall-out a highly tentative business. About all one can hope to do is to define what *might* happen. Assuming a 15 megaton superbomb burst close to the ground the author has made the following estimates for the fall-out ellipses:

Time (after burst)	Area	Average Intensity (gamma radiation)
1 hr	250 sq. mi.	2500 roentgen/hr
3 hr	1200 sq. mi.	200 roentgen/hr
6 hr	4000 sq. mi.	30 roentgen/hr

It must be remembered that this fall-out will occur downwind. Upwind for 15 miles and sideways for 20 miles the fall-out should not be lethal. Integration of the dosage rate for the 4,000 square mile area leads to the conclusion that people will receive a serious to lethal dose in the first day. The area and hazard represents a conservative calculation. The radioactive hazard is truly immense. The explosion of 50 superbombs could blanket the entire N.E. USA in a serious to lethal radioactive fog.

A schedule for the effects of external radiation upon a man in the open is as follows:

Roentgen Dose*	Effect on Man
50 to 100 r	Few per cent casualties
150 to 200 r	50 percent casualties
200 to 300 r	100% casualties plus some mortality
400 to 500 r	50% mortality
700 r	close to 100% mortality

Applying the 500 roentgen criterion to the fall-out pattern at 1 hour after the bomb burst it is clear that an individual within the 2500 r/hr area would accumulate this dose in 12 minutes. At 3 hours a person much farther downwind might be exposed to a very serious dose. Here we must consider the relation between the intensity of the radioactivity and the time of exposure. Fortunately, the process of radioactive decay reduces the intensity quite sharply as time goes by. However, one must seek shelter as a protection against the radiation if caught in a near-lethal area. That is, unless some means is provided to direct downwind populations out of the path of the fall-out.

Just what kind of shelter protection is required? The answer depends upon how close to ground zero one locates the shelter since blast then becomes the criterion. Assuming, however, that the shelter is located beyond the range of primary blast, the radiation shielding requirements are as follows:

Reduction Factor	10	50	100	1000
Inches of concrete	6	11	13	19
or Inches of packed soil	11	18	21	30

The relatively small thicknesses of concrete or earth shielding needed to reduce the incident radiation to one-hundredth of its topside value may surprise the layman. Exponential absorption of gamma radiation accounts for the fact that a foot and a half of hard packed soil can reduce an intensity of 2500 r/hr to 50 r/hr. Thirty inches of soil cuts this intensity down to 2.5 r/hr which can be regarded as acceptable for survival in a shelter.

*   *   *

---

*Delivered in a period of less than 1 hour. If delivered over several days the required effect dosage is doubled.

In addition to the external radiation hazard there is the enigma of ingested or inhaled radioactive debris. Judging from the brief reports issued by the Atomic Energy Commission on the medical histories of the Marshall Islanders exposed to fall-out, the internal radiation hazard may be less serious than generally believed. However, data on the Japanese survivors does not make for complacency on the significance of radioactive material taken into the body.

From the foregoing description it can be readily appreciated that fall-out presents civil defense with potentially greater perils than those of heat and blast. Blast can be readily felt as can heat and they both come in a flash. Radioactivity, on the other hand, cannot be felt and possesses all the terror of the unknown. It is something which evokes revulsion and helplessness—like a bubonic plague.

Having been already weighted down by the incubus of the immense primary effects of the superbomb, civil defense planners may well feel that they are condemned to a labor of a Sisyphus. Unless the challenge of the fall-out is met head-on, the very burden may kill civil defense in this country. Those who oppose the FCDA policy of evacuation may use fall-out as the excuse for rejecting the recommendation. They can (and with partial justification) point out that the mass removal of metropolitan population to the suburbs or open country may be like jumping from the frying pan into the fire.

Civil defense must reckon with the hazards of fall-out, but it would be utterly disastrous if it abandoned its policy of evacuation at this time. No one is going to come up with a perfect civil defense plan. As long as we have such huge agglomerations of people on a few bits of territory there can be no perfect civil defense. There will always have to be the element of the calculated risk to civil defense just as there is for the soldier at the front line. What civil defense must do is to acknowledge certain risks in its planning.

First things must come first in any good C.D. plan. This means that one must deal with certain primary effects of superweapons. The great circles of heat and blast effects for weapons in the 10 to 40 megaton range are simply too large to neglect. They are predictable whereas fall-out is not. Furthermore, *it makes no sense whatsoever to plan for a secondary hazard if you fail to survive the primary effects of the bomb.*

The Federal Civil Defense Administration has assessed the true significance of the circles of damage from H-bombs and has boldly put forth an evacuation policy to deploy urbanites to the suburbs beyond the 12 to 14 mile radius of the city. This policy has yet to be implemented but cities are going ahead with their plans. Milwaukee, Washington, Seattle, and Detroit are a few that have reached an advanced state of planning.

Consideration of the fall-out ellipses must be made and the C.D. plan of each city modified accordingly. One thing is perfectly clear—evacuation even on present plans does provide a large measure of protection for people since *distance is still the best defense against the bomb.* Those who are deployed upwind or laterally will escape the lethal fall-out. In general, viewed as a problem in geometry, it is the area of the evacuation circle which is intersected by the fall-out ellipse which requires special planning. It will be argued that since we have so many unknowns it is futile to plan. For example, the wind may be in a contrary direction. However, local data on upper air winds often show a pattern of prevailing winds which may indicate preferential evacuation in upwind directions. Moreover, when an alert comes civil defense directors can make an immediate prognostication of the probable fallout area based on current weather data.

The alert is the great and vexing problem in evacuation. It is often pointed out that there may be no warning whatsoever. This is possible but it certainly does not apply to all cities. Moreover, it flies in the face of probability. There may, in fact, be a long period of alert—a strategical alert of days, weeks, or even months.

Dr. Vannevar Bush, in his testimony before the Riehlman subcommittee, pointed out that conventional warfare might precede nuclear warfare due to the initial respect which each nation might have for the other's nuclear capability. Such a period of tension would make possible gradual decentralization of urban populations in anticipation of a nuclear attack. Time would also be given, plus the incentive of imminent attack, to prepare peripheral shelters to accommodate evacuees.

Besides the strategical alert period accompanying Dr. Bush's concept, there is a shorter term alert which would be concommitant with Soviet aggression into zones to which we guaranteed military support. Should the Soviets strike in such an area they might well limit themselves to conventional weapons, and it would then be up to the President to issue the fateful ultimatum for "retaliation in advance." Any such ultimatum would be, if sanity prevailed, accompanied by a nation-wide evacuation order.

The announcement that a Distant Early Warning line of radar posts is to be established in the far north provides hope that before long a 4 hour tactical alert will be assured for continental cities. Four hours is sufficient for most cities to deploy their populations to relatively safe sites provided evacuation plans are perfected and test drills are inaugurated.

Progress is being made in civil defense despite the vertiginous, almost exponential, rise in the hazards faced. The year 1954 may well mark the turning point in our C.D. activities. One very favorable index is that more and more top advisors in the government are becoming serious about civil defense. More and more, it is becoming clear that the security of the home base is of paramount importance. In this security, civil defense must assume a high priority.

The Federal Civil Defense Administration must be permitted access to classified data about fall-out. Furthermore, the agency must be able to translate these data or "sanitize" them so that a realistic picture of the radioactive hazard can be given to the American people. Local communities should not be left to plan in the dark nor should they be put in the position of planning on the basis of newspaper reports.

In the final analysis the solution to civil defense problems will not come from Washington. It is in the local communities that they will find a solution. The federal government can help with solid planning data, with technical assistance, and with financial support. But the real burden of the work must be shouldered by the people who reside in the cities. However, the new peril from radioactive fall-out is more than just a threat to civil defense—it is a peril to humanity.

41. Atomic Energy Commission Meeting 1376, May 21, 1958

2:15 P.M., Wednesday, May 21, 1958, Room A-410
Germantown, Maryland

DELETED

7. Suspension of Tests

The Commissioners discussed with the General Manager and General Starbird the question of suspension of nuclear weapons tests in connection with its consideration during the forthcoming meeting of President Eisenhower and Prime Minister MacMillan.

The Commissioners agreed that in view of the possibility of establishment of a moratorium on weapons testing, the Commission should be informed of the advantages and disadvantages of underground weapons testing in the event that it is considered desirable to propose continuation of such tests.

This document appeared in the United States Atomic Energy Commission Records, United States Department of Energy, Germantown, Maryland.

The Commissioners requested General Starbird to arrange a meeting of the Commission with Messrs. Bradbury, Teller, and others on May 28 to discuss the question of underground shots.

W. B. McCool
Secretary

42. Atomic Energy Commission Meeting 1377, May 28, 1958

10:15 A.M., Wednesday, May 28, 1958, Room A-410, Germantown, Maryland

1. Weapons Test Limitations

Mr. Graham reviewed with the representatives of the AEC weapons laboratories a number of events which had led the Commission to request a meeting with them to discuss the question of weapons test limitations. He pointed out that the subject of weapons tests had been considered both by the General Advisory Committee and the Advisory Committee on Biology and Medicine at their recent meetings. In addition, the current Congressional authorization hearings were concerned, in part, with funds to be authorized for the weapons test program. Mr. Graham said that all of these events had made it evident that the question of future weapons testing was still far from being resolved, and therefore he and other Commissioners believed it would be helpful if a full discussion of the subject could be held with representatives from the weapons laboratories. General Starbird added that the Laboratory Directors had in particular been requested to consider what technical problems were involved and what limitations would result from a decision to test underground only.

Mr. Teller began the discussion by stating that scientists at Livermore Laboratory had concluded that nearly all the information needed by the U.S. about nuclear explosions could be obtained from underground tests, and that underground weapons testing is more easily carried out than testing above ground. He said that he would be in favor of conducting future

This document appeared in the United States Atomic Energy Commission Records, United States Department of Energy, Germantown, Maryland.

weapons tests underground regardless of whether an international test limitation should be agreed upon. He added, however, that before it could be known with certainty that weapons in the megaton range could be successfully tested underground, there would have to be a series of tests gradually increasing in size to the megaton range. In addition, if all tests were conducted underground, there would be no opportunity to test weapons effects or to proof-test anti-ICBM missile systems. Therefore, he recommended that it would be desirable to have some above-ground testing each year but to limit each country to tests placing a maximum DELETED of fissionable material in the atmosphere per year. This amount of above-ground testing, he said, would permit all the diagnostic weapons experimentation necessary for the weapons program. If the maximum amount of fissionable material which could be put into the atmosphere each year by each country were limited DELETED it would still be possible to conduct valuable above-ground tests, although the amount of information gained would be less and the results would be limited.

Mr. Brown discussed with the Commissioners the three major types of measurements for a weapons test and how accurate these measurements would be if applied to an underground detonation. Yield measurement for the Rainier shot, he said, had a margin of error DELETED DELETED but with additional experimentation, this margin could be brought down DELETED. Mr. Teller remarked that weapons yields could not be quite as accurately determined underground as above-ground. However, diagnostic details from direct measurements can be obtained more easily and more accurately underground than above-ground, he said.

Mr. Brown said that, in his opinion, radio-chemistry measurement techniques for underground tests, although not as accurate as yield measurements, are adequate. Reaction history measurements, he said, can actually be more accurate underground than above-ground. Mr. Brown said he agreed with Mr. Teller that all necessary above-ground experiments could be conducted within the limitation DELETED DELETED. . . . to keep within this limitation, he said it would probably be necessary to substitute clean weapons for normal weapons shots. Mr. Brown said special weapons types would have to be developed for effects testing and for anti-missile missile warheads to be detonated underground.

Mr. Brown said the best information available to him indicated that weapons DELETED could be tested safely under-ground but, as Mr. Teller had stated, the size of the tests would have to be detonated deeper and in much harder rock, and there is uncertainty about the effect of such detonations since the transmission of shock would be more direct.

Mr. Sewell reviewed with the Commissioners the advantages of an underground test program over the present series of yearly above-ground test programs. Greater flexibility of scheduling tests and therefore more rapid progress in developing new weapons would be possible if the laboratories were not restricted to waiting for only one test series each year. With an underground test program, tests could be conducted periodically throughout the year whenever a weapon under development reached the point where certain test experiments needed to be conducted. More radical weapons designs could be tested because a laboratory would know that if a particular test failed, a year would not have to elapse before it could conduct another test experiment. This fact could also lead to more rapid progress in weapons development. Costly, full-scale test operations such as are conducted in alternate years at the Pacific Proving Grounds could be eliminated. Tests would no longer be dependent upon weather conditions, thus saving additional time and money now consumed in conducting above-ground tests. The cost of digging the tunnel for the Rainier shot cost no more than a five-hundred foot tower for a test device, and the cost of digging new tunnels out from a main one is about one-fourth the cost of the original tunnel Mr. Sewell said. In addition, public opposition to the tests because of the fallout danger could be eliminated by under-ground testing.

Mr. Teller stated that conducting all future U.S. weapons tests underground would interfere somewhat with Project PLOWSHARE, the program for developing peaceful uses of nuclear explosive devices.

In response to a question by Mr. Fields, Mr. Sewell said as many DELETED bombs could be detonated in a single mountain.

Mr. Bradbury commented that although the Livermore scientists may be correct in believing that nearly all necessary weapons information could be obtained from underground testing, if actual experience proves this to be incorrect, the U.S. would be placed in a very difficult position. He pointed out that much of this work on megaton weapons now involved detailed improvements of the weapons and that with only underground testing there would be a greater risk of being misled about what the tests actually demonstrated. Mr. Bradbury said he took the position that a diversified testing program using balloons, towers, and underground tunnels would be preferable to a complete underground test program.

Mr. Teller remarked that data on a weapons test is most needed when that device misfires, and that the diagnostics of such a test could be done as well or better if the shot were underground rather than above-ground.

General Starbird pointed out that estimates regarding a major underground test program are being based to a large extent upon extrapolations from the Rainier shot. He said he concurred with Messrs. Bradbury and Graves in their view that the U.S. would be faced with a serious military problem if it agreed to underground testing for all weapons, and then discovered that the larger weapons could not be satisfactorily tested underground. He added that the public could not be expected to raise questions about the safety of even underground weapons tests, and the danger that the radio-activity created by the detonation might sometime contaminate water and plant life.

Mr. Graves said he favored the use of balloons for testing weapons devices and that this valuable testing technique would be lost if all aboveground tests were halted.

Mr. Teller stated the belief that there would be no chance that radio-activity from underground tests could contaminate animals or humans, and said he believed one could be overly cautious about questioning the feasibility of underground testing.

Mr. Vance referred to a statement by the GAC at their 58th meeting to the effect that the Commission must not wait too long in proposing an intermediate position between unlimited testing and a ban on all weapons testing. Alternatives suggested by Mr. Vance included: (1) A limitation on the amount of radioactivity released into the atmosphere; (2) Supervision by the U.N. of all weapons tests; and (3) Complete underground testing of all weapons. Mr. Vance said he favored the underground testing position because it would be the simplest to establish and because it would end completely any radioactive contamination of the atmosphere from weapons detonations. He said he considered that public concern with the question makes it an important consideration, even though the Commission recognized that the danger at present is negligible. Mr. Vance said the question he wished the representatives of the laboratories to discuss is what intermediate position would enable the U.S. to continue its weapons development program with the greatest freedom. Mr. Bradbury said it would be difficult to select the best alternative since each one would interfere with the development which is possible under unlimited testing.

Mr. Teller said he believed the most desirable alternative would be to restrict above-ground testing by each country to DELETED fission products released into the atmosphere each year, and to require that all other tests be conducted underground. Such a position would have a number of advantages, he said. It would permit the development of anti-missile missiles, which probably would be impossible if all testing had to be conducted underground. It would also make it quite difficult for nations which do not now have weapons capability to develop weapons. They would be forced to develop both very small weapons for underground testing and clean devices for above-ground testing. Finally, an agreement to limit the amount of fissionable material released into the atmosphere would necessitate establishing a mutual inspection system, thus giving U.S. observers an opportunity to gain first-hand information about the Soviet weapons program.

Mr. Teller said he would be reluctant to have the U.S. accept more stringent test limitations that the DELETED fission yield per year above ground, and unlimited underground testing.

Mr. Vance observed that Mr. Teller's arguments were quite logical; however, the question of limiting nuclear tests would not necessarily be resolved on a logical or strictly scientific basis. He cited the instance of Mr. Hans Bethe's recent statements to the effect that all nuclear testing should be halted as a first step in ending the arms race and reducing world tensions. These, he said, are political judgments, not scientific ones, although they are made by a scientist.

Mr. Floberg asked Mr. Teller how possible violations of limitations such as he recommended would be handled. Mr. Teller replied that he would not be too concerned if the violation were in the range DELETED of fission yield put into the atmosphere in a single year, DELETED. Violations substantially greater than this, however, would be a cause for serious concern, he said. He pointed out that all countries carrying out an atmospheric sampling program would know when serious violations had occurred. General Starbird remarked that continued violations of such a limitation might cause neutralist countries to conclude that the limitation was not effective and to demand a complete cessation of all types of weapons testing.

Mr. Strauss raised the question of how complete weapons systems could be proof-tested underground. Mr. Bradbury said he was not convinced that a final proof-test of a missile and warhead was absolutely necessary if the two could be adequately tested separately. Mr. Teller said a great amount of experimentation was still required for an anti-ICBM missile warhead and that some above-ground testing would be necessary.

Mr. Strauss then inquired whether fallout measurement techniques are accurate enough to determine precisely whether DELETED DELETED fission yield had been placed in the atmosphere by a particular country. Mr. Teller replied that this would be possible only if there were mutual inspection within the countries where the tests occurred.

The question of identifying from which country particular fallout originated was then discussed by Mr. Strauss and Mr. Teller. Mr. Strauss postulated that a country might detonate a number of weapons at extremely high altitudes, with the fallout not being detected until a year or more later. He asked whether it would be possible then to accurately determine which country was responsible for those tests. Mr. Teller said that with satellite counters it would be possible to detect almost immediately when high altitude shots have been detonated, but that the country carrying out the test could not be positively identified in such cases. He added, however, that scientists could probably make an intelligent guess about the origin of the shots.

Mr. Vance left the meeting briefly during the above discussion. The Commissioners returned to the question of the necessity of proof-testing complete weapons systems, and Mr. Graham inquired whether it would be satisfactory to test missile systems using chemical rather than nuclear warheads. Mr. Teller said it would be possible to test a missile system without the nuclear warhead and determine the performance of such things as the timing of the detonation device, but of course, such a test would not provide any information on the effects of a nuclear explosion. Mr. Floberg said he believed there have been many examples in the past of the importance of conducting tests of complete weapons systems, such as submarine torpedos. He said he was unconvinced that it would be unnecessary to carry out proof-tests of the complete missile system. The Psychological factor of working with missiles equipped with nuclear warheads was important in itself, he said.

Mr. Fields then discussed with Mr. Teller and Mr. Graves when it would be possible to carry out the first diagnostic tests of an anti-ICBM missile. Mr. Brown said the Air Force had mentioned 1960 as the date when such tests might be held; however, the weapons laboratories do not yet know what device might be used. Mr. Teller said the DELETED may be a basis for developing such a weapon but that much more experimentation is required before it could be perfected. Following the detonation DELETED at Operation HARDTACK, it might be possible to conduct further test of it in 1959. Mr. Graves said he believed that it probably would be 1960 before an anti-ICBM device would be ready for diagnostic testing.

Mr. Strauss concluded the discussion by expressing the Commissioners' appreciation to Messrs. Teller, Bradbury, Graves, Brown, and Sewell for an enlightening discussion of the test limitations questions.

2. Proposed Additional Shots for Operation HARDTACK

General Starbird reported that it was planned to complete the Johnston Island test by September 1, 1958, and that Admiral Stump wishes to announce the danger area for that shot one month in advance. With fair weather conditions, he said he hoped the tests at Bikini and Eniwetok could be concluded by the middle of August. However, General Starbird said the Department of State has said that when the announcement of the new danger area around Johnston Island is made, the closing of the other danger area should also be announced.

General Starbird then reviewed with the Commissioners five additional shots which the DOE had requested be conducted during Operation HARDTACK.

DELETED

The Commissioners discussed this matter briefly and deferred a decision until the afternoon meeting.

DELETED

8. AEC 784/40–Proposed Legislation in Connection with the Indemnity Program

Mr. Fields presented to the Commissioners proposed amendments to the Price-Anderson Indemnity Act and inquired whether the Commissioners would object to his discussing the proposed amendments with BOB officials. The Commissioners indicated they would not object to such discussions. They made no decision at that time on the proposed amendments.

W. B. McCool
Secretary

43. W. K. Wyant, Jr., "50,000 Baby Teeth," *The Nation*, June 13, 1959

Ordinarily, a group that called itself the Greater St. Louis Citizens Committee for Nuclear Information would not be expected to last for any great period of time. Mortality among earnest and well-meaning organizations has been as great here as elsewhere, and the waters of the Mississippi have rolled over many such. Yet the committee now is striding vigorously

This essay appeared in *The Nation*, June 13, 1959.

into its second year. It has turned out to be an unusually happy union of scientific knowledge and civic leadership. Lay members are not afraid of Ph.D.s and M.D.s, the good doctors are neither fearful nor contemptuous of the laymen, and both doctors and laymen are unafraid of the United States Government.

At the outset, CNI—as the committee is called locally—decided it would not be an "action" group—that is to say, it would take no position for or against testing of nuclear weapons, even though the sentiment of the organizing spirits was clearly and outspokenly against. The view prevailed that what really was needed was information. It was felt that too many people—the politicians, the military and the oracles speaking "ex cathedra" from the Atomic Energy Commission—were taking decisive attitudes on the basis of indecisive information, or none.

This stand, useful in itself, also served the practical purpose of broadening the base of support, with the consequence that the CNI leadership now reflects a wide spectrum of opinion.

But while persisting in its "non-action" policy, the committee during its first year waged such a valiant fight on the information front that strontium-90 is now a household word in St. Louis. On the average of every other day, a member of the speakers' stable makes a talk on the subject. The CNI's monthly bulletin's circulation has jumped from 500 to 2,500. And the committee's drive to collect 50,000 baby teeth, to be analyzed for strontium-90, got headlines throughout the world press.

A few months ago, CNI's first anniversary meeting took place at the Union Avenue Christian Church, where the organization had been founded. Present was a cross-section of the medical and scientific elite of the city, along with lawyers, ministers and other civic leaders. Morale was high. There was evident a fine rapport, an atmosphere of comradeship and understanding, between scientists and laymen. One reason for this was that much had been accomplished with little.

Dr. Alfred S. Schwartz, assistant professor of clinical pediatrics at Washington University School of Medicine, the committee's vice president and treasurer, reported $5,077 had been spent during the year, leaving a balance of $469. Membership was more than 500. Tribute was paid to the women volunteer workers, many of them the wives of physicians and scientists.

Specifically, what had the committee done in its first year? There was a good deal besides the fact that it now had an office on the second floor of an old, converted red-brick house on West Pine Boulevard, with a number in the telephone book, a constantly ringing telephone, and a dedicated staff secretary to answer phone and mail—Mrs. Edward C. Roberts, who started as a part-time paid worker and now is a full-time volunteer.

"50,000 Baby Teeth"

1. A Speaker's Bureau, consisting of seventeen men and three women, has been organized to offer scientists as speakers to organizations in the St. Louis area. The speakers are all either M.D.s or Ph.D.s, capable of dealing with technical aspects of the fallout controversy. Since last October, some seventy church, parent, fraternal and business groups have heard addresses. Three organizers of the committee—Barry Commoner, professor of botany at Washington University, Dr. Walter C. Bauer, instructor in surgical pathology at the university's School of Medicine, and John M. Fowler, Washington University physicist—have made the knife-and-fork circuit a way of life. Fowler estimated they spend about a third of their time at it.

2. *Nuclear Information*, CNI's monthly publication, was started last fall to report new scientific facts in lay language. The March issue was entitled "Strontium-90 and Common Foods." It served to point out that the public had heard much about radioactivity in milk, but little about radioactivity in other foods. An article told of the three-year survey of wheat samples from Minnesota and the Dakotas.

3. CNI has sponsored two series of five seminars each on radiation for the scientific and medical community, and three public meetings at which qualified people discussed the consequences of radiation and fallout.

4. At the request of the St. Louis Dairy Council, CNI scientists last January issued a statement discussing the potential hazard from strontium-90 in milk (the local milk supply has been showing the highest strontium-90 concentration of 10 cities surveyed monthly by the United States Public Health Service—and nobody knows why). The statement said, in effect, that the harm that might result to children from milk containing radioactive substances could not yet be assayed with certainty. It emphasized that milk, as an essential food, must not be eliminated from the diet, and called for research on ways to lower its strontium-90 content.

5. The Baby Tooth Survey was started last December as a ten-year scientific project. Directed by Dr. Louise Reiss, an internist, the campaign has the cooperation of the School of Dentistry at both Washington and St. Louis universities. It has also served as an excellent device for calling attention to the more dreadful implications of nuclear fallout.

There was deep concern in St. Louis about nuclear-weapons testing for several years before CNI was organized. One of Washington University's most widely known scientists, former Chancellor Arthur H. Compton, Nobel Prize-winner and key figure in developing the A-bomb in World War II, has maintained a friendly attitude toward government atomic policies. But younger men in the physics department, headed by Edward U. Condon, former U.S. Bureau of Standards, chief, were openly hostile. In the spring of 1957, Linus C. Pauling of the California Institute of Technology, also a Nobel Prize-winner, attacked atom bomb testing in an eloquent address before Washington University's faculty and student body, warning of the price that might have to be paid by generations yet unborn. Later, in Condon's office, Pauling drew up a petition that called for immediate action to halt testing by international agreement.

Condon, Commoner and three other faculty members were among the twenty-seven signers of this petition; by the time Pauling presented it to the U.N. Secretary-General in January, 1958, more than 9,000 scientists from forty-three nations had subscribed. The scientific community at St. Louis University, a Catholic institution, had a share in the effort.

The formation of CNI followed several years of public needling of the AEC by Fowler, Commoner, Dr. Bauer, T. Alexander Pond of Washington University's physics department, and others. Some were horrified at the attacks made on Adlai Stevenson when he tried to debate the fallout question in the 1956 Presidential campaign.

St. Louis women, an enlightened and militant breed, got into the conflict early. A pioneer group calling itself "Eves Against Atoms," headed by Mrs. Thomas B. Sherman, wife of a *Post-Dispatch* editor, took a stand against testing. However, the CNI traces its origin directly to a meeting which took place in March of last year at the apartment of Mrs. George Gellhorn, a tireless and effective worker for civic causes in city, state and nation for sixty years. Now eighty, a widow since 1936 and the mother of writer Martha Gellhorn, Mrs. Gellhorn is a woman possessed of beauty, charm and sharp political savvy. Her assistance in any campaign to get something done in St. Louis is the rough equivalent of six Marine battalions.

From the start, CNI has had strong moral and religious overtones, sharpened by a kind of cheerful "let's look at the facts" iconoclasm. Like Mrs. Gellhorn, the Commoners—Barry and his wife, Gloria, committee vice president—are members of the St. Louis Ethical Society. So is Alexander S. Langsdorf, dean emeritus of Washington University's School of Engineering and CNI president. Fowler and Dr. Bauer are Quakers. At the initial meeting in the Gellhorn apartment, in addition to Commoner and Fowler, was the Rev. Ralph C. Abele, then head of the Metropolitan Church Federation. They decided to form a permanent body.

The next step was a meeting of about thirty people which took place several weeks later at the home of Mrs. Ernest W. Stix, another civic leader. Attending were university people, members of social-action groups, priests, ministers, housewives. The sequel was the founders' meeting, April 21, 1958.

While not all the CNI prime movers took a position against testing, the major figures did; and it is quite apparent that the group as a whole is motivated by deep misgivings as to the biological and moral implications of the weapons race. Nevertheless, the decision was to take no official stand and to concentrate instead on information.

Scientists with CNI have served as a Greek chorus to amplify, comment on, question, explain and criticize the thunderous pronunciamentos of the protagonists in Washington. Each new gobbet of official information is analyzed and put into context. Research papers are seized on, dissected and translated into lay language. The press is kept alerted to its duty. Any inaccuracies, omission or glossings-over in official statements are promptly challenged.

The Baby Tooth Survey has given the organization an objective that should hold it together for some time. The idea was suggested last summer by Dr. Schwartz, who cited an article in the British scientific publication *Nature* written by Dr. Herman M. Kalckar, a biochemist at Johns Hopkins University. CNI became the first group anywhere, apparently to initiate a large collection of "milk" teeth to be analyzed for radioactivity.

It was characteristic of the committee that it planned the project carefully. Before any announcement was made, its scope and purpose had the approval of the deans of both the local schools of dentistry; and Dr. John T. Bird, assistant dean of Washington University's school had agreed to head a group of dentists to examine and classify the teeth. Dr. E. S. Khalifah, editor of the *Journal* of the Missouri State Dental Association, gave his enthusiastic support. With the help of Martin Quigley, a public-relations adviser who is on the CNI board, the announcement, when it came, was deliberately designed to softpedal the "human interest" angle of the story and to stress sober scientific objectives. It said in part:

The importance of an immediate collection of deciduous, or baby, teeth lies in the fact that teeth now being shed by children represent an irreplaceable source of scientific information about the absorption of strontium-90 in the human body. Beginning about ten years ago, strontium-90 from nuclear-test fallout began to reach the earth and to contaminate human food.

Deciduous teeth now being shed were formed from the minerals present in food eaten by mothers and infants during the period 1948 to 1953—the first few years of the fallout era—and therefore represent invaluable baseline information with which analysis of later teeth and bones can be compared. Unless a collection of deciduous teeth is started immediately, scientists will lose the chance to learn how much strontium-90 human beings absorbed during the first years of the atomic age.

Strontium-90 present in food accumulates in bones and teeth; milk is the main food source of strontium-90. In sufficient amounts the radiation from strontium-90 may cause harmful effects, including bone tumors and other forms of cancer. . . .

The reaction to the announcement was instantaneous and world-wide: There were letters from New York, Hawaii, Calcutta, California, Spain. Some mail from afar included teeth, although the committee wants specimens only from the St. Louis area.

Dr. Louise Reiss and her assistants have had extraordinary success in getting local schools—public, private and parochial—to help in the teeth collecting. Some 250,000 forms have been distributed to reach all lower-grade students. The Council of Catholic Women, the public libraries, the dental societies and city dental clinics have been of great assistance. At present, baby teeth are reaching the little office on West Pine Boulevard at the rate of about fifty a day. The stockpile is still short of 10,000, although the estimated annual "fallout" of baby teeth in the St. Louis area is half-a-million. When enough have been accumulated, they will be classified, ground up and analyzed at the Washington University School of Dentistry. Mothers who request an individual report on their children's teeth are doomed to disappointment.

Summing up the first year, the CNI, adopting a "conservative" approach, has set a pattern already being followed by other communities. What official bodies will not do for them, citizens are seeking to do for themselves.

---

44. Treaty Banning Nuclear Weapon Tests in the Atmosphere, in Outer Space, and Under Water, August 5, 1963

Signed at Moscow August 5, 1963
Ratification advised by U.S. Senate September 24, 1963
Ratified by U.S. President October 7, 1963
U.S. ratification deposited at Washington, London, and Moscow October 10, 1963
Proclaimed by U.S. President October 10, 1963
Entered into force October 10, 1963

This document appeared in United States Arms Control and Disarmament Agency, *Arms Control and Disarmament Agreements* (Washington, D.C.: Government Printing Office, 1982), pp. 41–43.

Treaty Banning Nuclear Weapon Tests

The Governments of the United States of America, the United Kingdom of Great Britain and Northern Ireland, and the Union of Soviet Socialist Republics, hereinafter referred to as the "Original Parties,"

Proclaiming as their principal aim the speediest possible achievement of an agreement on general and complete disarmament under strict international control in accordance with the objectives of the United Nations which would put an end to the armaments race and eliminate the incentive to the production and testing of all kinds of weapons, including nuclear weapons.

Seeking to achieve the discontinuance of all test explosions of nuclear weapons for all time, determined to continue negotiations to this end, and desiring to put an end to the contamination of man's environment by radioactive substances.

Have agreed as follows:

Article I

1. Each of the Parties to this Treaty undertakes to prohibit, to prevent, and not to carry out any nuclear weapon test explosion, or any other nuclear explosion, at any place under its jurisdiction or control:

(a) in the atmosphere; beyond its limits, including outer space; or under water, including territorial waters or high seas; or

(b) in any other environment if such explosion causes radioactive debris to be present outside the territorial limits of the State under whose jurisdiction or control such explosion is conducted. It is understood in this connection that the provisions of this subparagraph are without prejudice to the conclusion of a treaty resulting in the permanent banning of all nuclear test explosions, including all such explosions underground, the conclusion of which, as the Parties have stated in the Preamble to this Treaty, they seek to achieve.

2. Each of the Parties to this Treaty undertakes furthermore to refrain from causing, encouraging, or in any way participating in, the carrying out of any nuclear weapon test explosion, or any other nuclear explosion anywhere which would take place in any of the environments described, or have the effect referred to, in paragraph 1 of this Article.

Article II

1. Any Party may propose amendments to this Treaty. The text of any proposed amendment shall be submitted to the Depositary Governments

which shall circulate it to all Parties to this Treaty. Thereafter, if requested to do so by one-third or more of the Parties, the Depositary Governments shall convene a conference, to which they shall invite all the Parties, to consider such amendment.

2. Any amendment to this Treaty must be approved by a majority of the votes of all the Parties to this Treaty, including the votes of all of the Original Parties. The amendment shall enter into force for all Parties upon the deposit of instruments of ratification by a majority of all the Parties, including the instruments of ratification of all of the Original Parties.

Article III

1. This Treaty shall be open to all States for signature. Any State which does not sign this Treaty before its entry into force in accordance with paragraph 3 of this Article may accede to it at any time.

2. This Treaty shall be subject to ratification by signatory States, Instruments of ratification and instruments of accession shall be deposited with the Governments of the Original Parties—the United States of America, the United Kingdom of Great Britain and Northern Ireland, and the Union of Soviet Socialist Republics—which are hereby designated the Depositary Governments.

3. This Treaty shall enter into force after its ratification by all the Original Parties and the deposit of their instruments of ratification.

4. For States whose instruments of ratification or accession are deposited subsequent to the entry into force of this Treaty, it shall enter into force on the date of the deposit of their instruments of ratification or accession.

5. The Depositary Governments shall promptly inform all signatory and acceding States of the date of each signature, the date of deposit of each instrument of ratification of and accession to this Treaty, the date of its entry into force, and the date of receipt of any requests for conferences or other notices.

6. This Treaty shall be registered by the Depositary Governments pursuant to Article 102 of the Charter of the United Nations.

Article IV

This Treaty shall be of unlimited duration.

Each Party shall in exercising its national sovereignty have the right to withdraw from the Treaty if it decides that extraordinary events, related to the subject matter of this Treaty, have jeopardized the supreme interests of its country. It shall give notice of such withdrawal to all other Parties to the Treaty three months in advance.

Article V

This Treaty, of which the English and Russian texts are equally authentic, shall be deposited in the archives of the Depositary Governments. Duly certified copies of this Treaty shall be transmitted by the Depositary Governments to the Governments of the signatory and acceding States.

IN WITNESS WHEREOF the undersigned, duly authorized, have signed this Treaty.

DONE in triplicate at the city of Moscow the fifth day of August, one thousand nine hundred and sixty-three.

LIMITED TEST BAN TREATY

For the Government of the United States of America	For the Government of the United Kingdom of Great Britain and Northern Ireland	For the Government of the Union of Soviet Socialist Republics

45. Treaty between the United States of America and the Union of Soviet Socialist Republics on the Limitation of Underground Nuclear Weapon Tests, July 3, 1974

The United States of America and the Union of Soviet Socialist Republics, hereinafter referred to as the Parties,

Declaring their intention to achieve at the earliest possible date the cessation of the nuclear arms race and to take effective measures toward reductions in strategic arms, nuclear disarmament, and general and complete disarmament under strict and effective international control,

Recalling the determination expressed by the Parties to the 1963 Treaty Banning Nuclear Weapon Tests in the Atmosphere, in Outer Space and Under Water in its Preamble to seek to achieve the discontinuance of all

This document appeared in United States Arms Control and Disarmament Agency, *Arms Control and Disarmament Agreements* (Washington, D.C.: Government Printing Office, 1982), pp. 167–68.

test explosions of nuclear weapons for all time, and to continue negotiations to this end,

Noting that adoption of measures for the further limitation of underground nuclear weapon tests would contribute to the achievement of these objectives and would meet the interests of strengthening peace and the further relaxation of international tension,

Reaffirming their adherence to the objectives and principles of the Treaty Banning Nuclear Weapon Tests in the Atmosphere, in Outer Space and Under Water and of the Treaty on the Non-Proliferation of Nuclear Weapons,

Have agreed as follows:

Article I

1. Each Party undertakes to prohibit, to prevent, and not to carry out any underground nuclear weapon test having a yield exceeding 150 kilotons at any place under its jurisdiction or control, beginning March 31, 1976.

2. Each Party shall continue their negotiations with a view toward achieving a solution to the problem of the cessation of all underground nuclear weapon tests.

Article II

1. For the purpose of providing assurance of compliance with the provisions of this Treaty, each Party shall use national technical means of verification at its disposal in a manner consistent with the generally recognized principles of international law.

2. Each Party undertakes not to interfere with the national technical means of verification of the other Party operating in accordance with paragraph 1 of this Article.

To promote the objectives and implementation of the provisions of this Treaty the Parties shall, as necessary, consult with each other, make inquiries and furnish information in response to such inquiries.

TESTS OF ARMS CONTROL AGREEMENTS

Article III

The provisions of this Treaty do not extend to underground nuclear explosions carried out by the Parties for peaceful purposes. Underground nuclear explosions for peaceful purposes shall be governed by an agreement which is to be negotiated and concluded by the Parties at the earliest possible time.

Article IV

This Treaty shall be subject to ratification in accordance with the constitutional procedures of each Party. This Treaty shall enter into force on the day of exchange of instruments of ratification.

Article V

1. This Treaty shall remain in force for a period of five years. Unless replaced earlier by an agreement in implementation of the objectives specified in paragraph 3 of Article I of this Treaty, it shall be extended for successive five-year periods unless either Party notifies the other of its termination no later than six months prior to the expiration of the Treaty. Before the expiration of this period the Parties may, as necessary, hold consultations to consider the situation relevant to the substance of this Treaty and to introduce possible amendments to the text of the Treaty.

2. Each Party shall, in exercising its national sovereignty, have the right to withdraw from this Treaty if it decides that extraordinary events related to the subject matter of this Treaty have jeopardized its supreme interests. It shall give notice of its decision to the other Party six months prior to withdrawal from this Treaty. Such notice shall include a statement of the extraordinary events the notifying Party regards as having jeopardized its supreme interests.

3. This Treaty shall be registered pursuant to Article 102 of the Charter of the United Nations.

Done at Moscow on July 3, 1974, in duplicate, in the English and Russian languages, both texts being equally authentic.

For the United States of America:
  RICHARD NIXON
    The President of the United States of America

For the Union of Soviet Socialist Republics:
  L. I. BREZHNEV
    General Secretary of the Central Committee of the CPSU

46. Protocol to the Treaty between the United States of America and the Union of Soviet Socialist Republics on the Limitation of Underground Nuclear Weapon Tests, July 3, 1974

The United States of America and the Union of Soviet Socialist Republics, hereinafter referred to as the Parties,

Have agreed to limit underground nuclear weapon tests,

Have agreed as follows:

1. For the Purpose of ensuring verification of compliance with the obligations of the Parties under the Treaty by national technical means, the Parties shall on the basis of reciprocity, exchange the following data:

a. The geographic coordinates of the boundaries of each test site and of the boundaries of the geophysically distinct testing areas therein.

b. Information on the geology of the testing areas of the sites (the rock characteristics of geological formations and the basic physical properties of the rock, i.e., density, seismic velocity, water saturation, porosity and depth of water table).

c. The geographic coordinates of underground nuclear weapon tests, after they have been conducted.

d. Yield, date, time, depth and coordinates for two nuclear weapon tests for calibration purposes from each geophysically distinct testing area where underground nuclear weapon tests have been and are to be conducted. In this connection the yield of such explosions for calibration purposes should be as near as possible to the limit defined in Article I of the Treaty and not less than one-tenth of that limit. In the case of testing areas where data are not available on two tests for calibration purposes, the data pertaining to one such test shall be exchanged, if available, and the data pertaining to the second test shall be exchanged as soon as possible after the second test having a yield in the above-mentioned range. The provision of this Protocol shall not require the Parties to conduct tests solely for calibration purposes.

2. The Parties agree that the exchange of data pursuant to subparagraphs a, b, and d of paragraph 1 shall be carried out simultaneously with the exchange of instruments of ratification of the Treaty, as provided in

This document appeared in United States Arms Control and Disarmament Agency, *Arms Control and Disarmament Agreements* (Washington, D.C.: Government Printing Office, 1982), pp. 169–70.

Article VI of the Treaty, having in mind that the Parties shall, on the basis of reciprocity, afford each other the opportunity to familiarize themselves with these data before the exchange of instruments of ratification.

3. Should a Party specify a new test site or testing area after the entry into force of the Treaty, the data called for by subparagraphs a and b of paragraph 1 shall be transmitted to the other Party in advance of use of that site or area. The data called for by subparagraph d of paragraph 1 shall also be transmitted in advance of use of that site or area if they are available; if they are not available, they shall be transmitted as soon as possible after they have been obtained by the transmitting Party.

4. The Parties agree that the test sites of each Party shall be located at places under its jurisdiction or control and that all nuclear weapon tests shall be conducted solely within the testing areas specified in accordance with paragraph 1.

For the purposes of the Treaty, all underground nuclear explosions at the specified test sites shall be considered nuclear weapon tests and shall be subject to all the provisions of the Treaty relating to nuclear weapon tests. The provisions of Article III of the Treaty apply to all underground nuclear explosions conducted outside of the specified test sites, and only to such explosions.

This Protocol shall be considered an integral part of the Treaty.

Done at Moscow on July 3, 1974.[1]

For the United States of America:
RICHARD NIXON
   The President of the United States of America.

For the Union of Soviet Socialist Republics:
L. I. BREZHNEV
   General Secretary of the Central Committee of the CPSU

---

1. The Treaty and protocol were signed at a ceremony in St. Vladimir Hall of the Grand Kremlin Palace on Wednesday, July 3, 1974.

VII

Deterrence

*I*n 1945 the United States became the first nation to use a nuclear weapon in time of war. Even then many scientists and policy makers realized that the terrible power of the nuclear age implied a terrible paradox, namely, that nuclear weapons were both a problem and a solution. The problem was that their enormous destructive power made them ill suited to any rational military or foreign policy; two rival nations with nuclear weapons could threaten each other but offer no real defense against an attack. The solution was that the nuclear threat could avert nuclear war precisely by its existence; under threat of mutual annihilation, two nations, like the two scorpions in a bottle in J. Robert Oppenheimer's metaphor, could destroy the other only at the risk of destroying themselves.

As early as 1946 Bernard Brodie and other strategic thinkers associated with the RAND Corporation, an air force think tank in California, realized that with nuclear weapons full-scale wars could be averted or deterred, but not really won. "Fear of substantial retaliation," wrote Brodie, would be the key to nuclear policy, which would not be to win wars, but to avert them. Moreover, the only rational purpose in stockpiling more nuclear weapons would be to ensure that they need not be used in time of war. The notion of an arms race as a means to peace became the paradoxical cornerstone of the doctrine of deterrence.

After 1949 the American atomic monopoly gave way to a Soviet-American arms race centered on the construction of thermonuclear weapons. In 1954 Secretary of State John Foster Dulles established as the basis of deterrence (of both war and Soviet aggression) the doctrine of massive retaliation, whereby the United States would maintain the peace by maintaining its ability to respond to a nuclear attack or any other form of aggression with an all-out nuclear attack upon the Soviet Union. Massive retaliation would be the American response to a Soviet first-strike attack on the United States and also to an attack on Europe and its new North Atlantic Treaty Organization (NATO). The delivery vehicles would be the B-52 bombers of the Strategic Air Command, and the weapons would be nuclear and thermonuclear.

In the late 1950s the development of intercontinental ballistic missiles reduced the time of delivery of strategic nuclear weapons to minutes, rather than hours. The winds of the Cold War blew vigorously, and Dulles had shifted the notion of deterrence from the deterrence of war to the deterrence of Soviet aggression. As the number of nuclear weapons multiplied, the danger of war by accident was compounded by an inflexible military doctrine that threatened nuclear response to nonnuclear but unspecified aggression. Thus was born the idea of limited nuclear war, laid down in quantitative and frightening terms by another RAND strategist, Herman Kahn, in his book On Thermonuclear War, published in 1960.

Kahn created a sensation by arguing that nuclear war might not mean the end of the world, as envisaged in On The Beach and other fictional scenarios of the 1950s. Rather, depending on how many weapons of what yield were used for how long a time, nuclear wars were distinguishable in their effects and postwar states; more than this, argued his critics, they were thinkable and thus winable. Brodie had argued that nuclear war was irrational; Kahn made it rational.

Until the 1960s limited accuracy and numbers of weapons meant that deterrence was limited to a "second strike" and "countervalue" exchange. That is, a nation would not strike first if it knew that its enemy would still retain the ability to strike back at its population centers and cities. Targeting cities, it was argued, was a stable and rational arms control measure in that it constituted only a strategy of retaliation; the intent to strike first would be indicated by targeting military forces, not cities.

In fact, a counterforce doctrine was emerging to provide a whole new theory of extended, not limited, deterrence. Under the air force-dominated SIOP targetting program, an increasing number of missiles in the 1960s were assigned to Soviet military targets. This shift in targeting was articulated in 1962 by Secretary of Defense Robert McNamara, who declared that the most rational policy would be to target military forces, not cities. "In particular, relatively weak national nuclear forces with enemy cities as their targets are not likely to be sufficient to perform even the function of deterrence," argued McNamara, since they would be vulnerable to attack. Thus by the late 1960s the number of available weapons, targeting strategy, and missile technology made it possible to target many times over both the military and civilian centers of the United States and the Soviet Union.

In the 1970s deterrence became complicated further by the development of another new technology, *Multiple Independently Targeted Reentry Vehicles*, or *MIRVs*. By placing as many as twelve warheads with independent computer guidance systems on the nosecone of a single missile, the ratio of available targets to each launcher was greatly increased. MIRV made the dream of an antiballistic missile defense hopeless, as well as expensive. It also made deterrence increasingly unstable. Until MIRV the doctrine of "mutual assured destruction," or MAD, was based on the fact that neither side could hope to destroy the launchers of the other in a first strike, so that a sufficient number of invulnerable second-strike launchers would always remain. MIRV technology made a first strike thinkable by targeting enemy forces many times over. The great success of the Strategic Arms Limitation Talks (SALT) of the 1970s was the 1972 agreement to limit ABM systems; the great failure was to allow the development of MIRVed missile launchers.

In 1979 extended deterrence was codified by the Carter administration in Presidential Directive 59, a top secret and still unpublished statement of American nuclear war-fighting policy. Coupled with Reagan administration talk of winning, limiting, and surviving a nuclear war by decapitating Soviet military command and control structures, PD 59 helped spawn a massive antinuclear movement in both Europe and America. An entire generation of young people joined concerned elders in questioning the entire basis of deterrence, a doctrine designed to avert war but widely viewed as merely a justification for an accelerating arms race. Led by Catholic bishops in America, the movement to question the entire philosophical and ethical justification for deterrence gathered momentum, especially as the December 1983 deadline for installing American cruise missiles in Western Europe approached.

The new sense of alarm about prospects for nuclear war was reflected most dramatically in Jonathan Schell's apocalyptic essay entitled The Fate of the Earth, published in 1982. Schell questioned the entire notion of deterrence but offered little in its place except the dream of world government and international control of atomic energy. The Soviet Union and the United States continued to negotiate arms reduction agreements and to stay within the rough limits of SALT II, an agreement limiting launchers that was signed in 1979 by President Carter and Premier Brezhnev but not ratified by the United States Senate. Yet deterrence remained the cornerstone of both American and Soviet nuclear policy. The concept of deterrence has undergone many changes since 1946, but no alternative short of nuclear war or unilateral disarmament has emerged. It remains the military theology of our times and, until a nuclear war occurs, will have worked.

The best survey of the theory of deterrence is Lawrence Freedman, The Evolution of Nuclear Strategy (New York: St. Martin's, 1981). On the history *of the RAND Corporation and its military intellectuals, see Fred Kaplan,* The Wizards of Armageddon *(New York: Simon and Schuster, 1983). A good brief view of deterrence as a necessary future is Michael Mandelbaum,* The Nuclear Future *(Ithaca and London: Cornell University Press, 1983). Jonathan Schell's* The Fate of the Earth *(New York: Alfred A. Knopf, 1982) is a radical critique of deterrence that should be read in conjunction with the more recent book by the Harvard Nuclear Study Group,* Living with Nuclear Weapons *(New York: Bantam Books, 1983), and, especially on the European perspective, Solly Zuckerman,* Nuclear Illusion and Reality *(New York: Vintage, 1983).*

Also critical of deterrence is Robert Jay Lifton and Richard Falk, Indefensible Weapons *(New York: Basic Books, 1982), and the more scholarly analysis by Louis René Beres,* Apocalypse: Nuclear Catastrophe in World Politics *(Chicago: University of Chicago Press, 1980). See also Beres, "Embracing Omnicide: President Reagan and the Strategic Mythmakers,"* The Hudson Review *(Spring 1983). The best speculation concerning what the failure of deterrence would mean is Arthur M. Katz,* Life after Nuclear War *(Cambridge, Mass.: Ballinger, 1982).*

---

47. Bernard Brodie, *The Absolute Weapon*, 1946

There is happily little disposition to believe that the atomic bomb by its mere existence and by the horror implicit in it "makes war impossible." In the sense that war is something not to be endured if any reasonable alternative remains, it has long been "impossible." But for that very reason we cannot hope that the bomb makes war impossible in the narrower sense of the word. Even without it the conditions of modern war should have been a sufficient deterrent but proved not to be such. If the atomic bomb can be used without fear of substantial retaliation in kind, it will clearly encourage aggression. So much the more reason, therefore, to take all possible steps to assure that multilateral possession of the bomb, should that prove inevitable, be attended by arrangements to make as nearly certain as possible that the aggressor who uses the bomb will have it used against him.

If such arrangements are made, the bomb cannot but prove in the net a

This essay appeared in Bernard Brodie, ed., *The Absolute Weapon: Atomic Power and World Order* (New York: Harcourt, Brace, 1946), pp. 75–77, 83–85, 106–7. Reprinted by permission of Harcourt Brace Jovanovich, Inc.

powerful inhibition to aggression. It would make relatively little difference if one power had more bombs and were better prepared to resist them than its opponent. It would in any case undergo incalculable destruction of life and property. It is clear that there existed in the thirties a deeper and probably more generalized revulsion against war than in any other era of history. Under those circumstances the breeding of a new war required a situation combining dictators of singular responsibility with a notion among them and their general staffs that aggression would be both successful and cheap. The possibility of irresponsible or desperate men again becoming rulers of powerful states cannot under the prevailing system of international politics be ruled out in the future. But it does seem possible to erase the idea—if not among madmen rulers then at least among their military supporters—this aggression will be cheap.

Thus, the first and most vital step in any American security program for the age of atomic bombs is to take measures to guarantee to ourselves in case of attack the possibility of retaliation in kind. The writer in making that statement is not for the moment concerned about who will *win* the next war in which atomic bombs are used. Thus far the chief purpose of our military establishment has been to win wars. From now on its chief purpose must be to avert them. It can have almost no other useful purpose.

Neither is the writer especially concerned with whether the guarantee of retaliation is based on national or international power. However, one cannot be unmindful of one obvious fact: for the period immediately ahead, we must evolve our plans with the knowledge that there is a vast difference between what a nation can do domestically of its own volition and on its own initiative and what it can do with respect to programs which depend on achieving agreement with other nations. Naturally, our domestic policies concerning the atomic bomb and the national defense generally should not be such as to prejudice real opportunities for achieving world security agreements of a worth-while sort. That is an important provision and may become a markedly restraining one.

Some means of international protection for those states which cannot protect themselves will remain as necessary in the future as it has been in the past.[1] Upon the security of such states our own security must ulti-

---

1. The argument has been made that once the middle or small powers have atomic bombs they will have restored to them the ability to resist effectively the aggressions of their great-power neighbors—an ability which otherwise has well-nigh disappeared. This is of course an interesting speculation on which no final answer is forthcoming. It is true that a small power, while admitting that it could not win a war against a great neighbor, could nevertheless threaten to use the bomb as a penalizing instrument if it were invaded. But it is also true that

mately depend. But only a great state which has taken the necessary steps to reduce its own direct vulnerability to atomic bomb attack is in a position to offer the necessary support. Reducing vulnerability is at least one way of reducing temptation to potential aggressors. And if the technological realities make reduction of vulnerability largely synonymous with reservation of striking power, that is a fact which must be faced. Under those circumstances any domestic measures which effectively guaranteed such preservation of striking power under attack would contribute to a more solid basis for the operation of an international security system.

Outlines of a Defense in the Atomic Age

What are the criteria by which we can appraise realistic military thinking in the age of atomic bombs? The burden of the answer will depend primarily on whether one accepts as true the several postulates presented and argued in the previous chapter. One might go further and say that since none of them is obviously untrue, no program of military preparedness which fails to consider the likelihood of their being true can be regarded as comprehensive or even reasonably adequate.

It is of course always possible that the world may see another major war in which the atomic bomb is not used. The awful menace to both parties of a reciprocal use of the bomb may prevent the resort to that weapon by either side, even if it does not prevent the outbreak of hostilities. But even so, the shadow of the atomic bomb would so govern the strategic and tactical dispositions of either side as to create a wholly novel form of war. The kind of spatial concentrations of force by which in the past great decisions have been achieved would be considered too risky. The whole economy of war would be affected, for even if the governments were willing to assume responsibility for keeping the urban populations in their homes, the spontaneous exodus of those populations from the cities might reach such proportions as to make it difficult to service the machines of war. The

---

the great-power aggressor could make counterthreats concerning its conduct while occupying the country which had used atomic bombs against it. It seems to this writer highly unlikely that a small power would dare threaten use of the bomb against a great neighbor which was sure to overrun it quickly once hostilities began. It seems, on the contrary, much more likely that Denmark's course in the second World War will be widely emulated if there is a third. The aggressor will not "atomize" a city occupied by its own troops, and the opposing belligerent will hesitate to destroy by such an unselective weapon the cities of an occupied friendly state.

conclusion is inescapable that war will be vastly different because of the atomic bomb whether or not the bomb is actually used.

But let us now consider the degree of probability inherent in each of the three main situations which might follow from a failure to prevent a major war. These three situations may be listed as follows:

(a)   a war fought without atomic bombs or other forms of radioactive energy;
(b)   a war in which atomic bombs were introduced only considerably after the outbreak of hostilities;
(c)   a war in which atomic bombs were used at or near the very outset of hostilities.

We are assuming that this hypothetical conflict occurs at a time when each of the opposing sides possesses at least the "know-how" of bomb production, a situation which, as argued in the previous chapter, approximates the realities to be expected not more than five to ten years hence.

Under such conditions the situation described under (a) above could obtain only as a result of a mutual fear of retaliation, perhaps supported by international instruments outlawing the bomb as a weapon of war. It would *not* be likely to result from the operation of an international system for the suppression of bomb production, since such a system would almost certainly not survive the outbreak of a major war. If such a system were in fact effective at the opening of hostilities, the situation resulting would be far more likely to fall under (b) than under (a), unless the war were very short. For the race to get the bomb would not be an even one, and the side which got it first in quantity would be under enormous temptation to use it before the opponent had it. Of course, it is more reasonable to assume that an international situation which had so far deteriorated as to permit the outbreak of a major war would have long since seen the collapse of whatever arrangements for bomb production control had previously been imposed, unless the conflict were indeed precipitated by an exercise of sanctions for the violation of such a control system.

Thus we see that a war in which atomic bombs are not used is more likely to occur if both sides have the bombs in quantity from the beginning than if neither side has it at the outset or if only one side has it.[2] But how likely is it to occur? Since the prime motive in refraining from using it

2. One can almost rule out too the possibility that war would break out between two great powers where both knew that only one of them had the bombs in quantity. It is one of the old maxims of power politics that *c'est une crime de faire la guerre sans compter sur la superiorité*, and certainly a monopoly of atomic bombs would be a sufficiently clear definition of superiority to dissuade the other side from accepting the gage of war unless directly attacked.

would be fear of retaliation, it is difficult to see why a fear of reciprocal use should be strong enough to prevent resort to the bomb without being strong enough to prevent the outbreak of war in the first place.

Of course, the bomb may act as a powerful deterrent to direct aggression against great powers without preventing the political crises out of which wars generally develop. In a world in which great wars become "inevitable" as a result of aggression by great powers upon weak neighbors, the bomb may easily have the contrary effect. Hitler made a good many bloodless gains by mere blackmail, in which he relied heavily on the too obvious horror of modern war among the great nations which might have opposed him earlier. A comparable kind of blackmail in the future may actually find its encouragement in the existence of the atomic bomb. Horror of its implications is not likely to be spared evenly, at least not in the form of overt expression. The result may be a series of *faits accomplis* eventuating in that final deterioration of international affairs in which war, however terrible, can no longer be avoided.

Those who have been predicting attacks of 15,000 atomic bombs and upward will no doubt look with jaundiced eye upon these speculations. For they will say that a country so struck will not merely be overwhelmed but for all practical purposes will vanish. Those areas not directly struck will be covered with clouds of radioactive dust under which all living beings will perish.

No doubt there is a possibility that initial attack can be so overwhelming as to void all opportunity of resistance or retaliation, regardless of the precautions taken in the target state. Not *all* eventualities can be provided against. But preparation to launch such an attack would have to be on so gigantic a scale as to eliminate all chances of surprise. Moreover, while there is perhaps little solace in the thought that the lethal effect of radioactivity is generally considerably delayed, the idea will not be lost on the aggressor. The more horrible the results of attack, the more he will be deterred by even a marginal chance of retaliation.

Finally, one can scarcely assume that the world will remain either long ignorant of or acquiescent in the accumulations of such vast stockpiles of atomic bombs. If existing international organization should prove inadequate to cope with the problem of controlling bomb production—and it would be premature to predict that it will prove inadequate, especially in view of the favorable official and public reception accorded the Board of Consultants' report of March 16, 1946—a runaway competition in such production would certainly bring new force into the picture. In this chapter and in the preceding one, the writer has been under no illusions concerning the adequacy of a purely military solution.

Concern with the efficiency of the national defenses is obviously inadequate in itself as an approach to the problem of the atomic bomb. In so far as such concern prevails over the more fundamental considerations of eliminating war or at least of reducing the chance of its recurrence, it clearly defeats its purpose. That has perhaps always been true, but it is a truth which is less escapable today than ever before. Nations can still save themselves by their own armed strength from subjugation, but not from a destruction so colossal as to involve complete ruin. Nevertheless, it also remains true that a nation which is as well girded for its own defense as is reasonably possible is not a tempting target to an aggressor. Such a nation is therefore better able to pursue actively that progressive improvement in world affairs by which alone it finds its true security.

48. John Foster Dulles on "Massive Retaliation," 1954

The Need for Long-Range Policies

This "long time" factor is of critical importance.

The Soviet Communists are planning for what they call "an entire historical era," and we should do the same. They seek, through many types of maneuvers, gradually to divide and weaken the free nations by over-extending them in efforts which, as Lenin put it, are "beyond their strength, so that they come to practical bankruptcy." Then, said Lenin, "our victory is assured." Then, said Stalin, will be "the moment for the decisive blow."

In the face of this strategy, measures cannot be judged adequate merely because they ward off an immediate danger. It is essential to do this, but it is also essential to do so without exhausting ourselves.

When the Eisenhower administration applied this test, we felt that some transformations were needed.

It is not sound military strategy permanently to commit U.S. land forces to Asia to a degree that leaves us no strategic reserves.

It is not sound economics, or good foreign policy, to support permanently other countries; for in the long run, that creates as much ill will as good will.

Also, it is not sound to become permanently committed to military expenditures so vast that they lead to "practical bankruptcy."

This essay appeared in John Foster Dulles, "The Evolution of Foreign Policy," *Department of State Bulletin*, 30 (January 25, 1962): 107–10.

Change was imperative to assure the stamina needed for permanent security. But it was equally imperative that change should be accompanied by understanding of our true purposes. Sudden and spectacular change had to be avoided. Otherwise, there might have been a panic among our friends and miscalculated aggression by our enemies. We can, I believe, make a good report in these respects.

We need allies and collective security. Our purpose is to make these relations more effective, less costly. This can be done by placing more reliance on deterrent power and less dependence on local defensive power.

This is accepted practice so far as local communities are concerned. We keep locks on our doors, but we do not have an armed guard in every home. We rely principally on a community security system so well equipped to punish any who break in and steal that, in fact, would-be aggressors are generally deterred. That is the modern way of getting maximum protection at a bearable cost.

What the Eisenhower administration seeks is a similar international security system. We want, for ourselves and the other free nations, a maximum deterrent at a bearable cost.

Local defense will always be important. But there is no local defense which alone will contain the mighty landpower of the Communist world. Local defenses must be reinforced by the further deterrent of massive retaliatory power. A potential aggressor must know that he cannot always prescribe battle conditions that suit him. Otherwise, for example, a potential aggressor, who is glutted with manpower, might be tempted to attack in confidence that resistance would be confined to manpower. He might be tempted to attack in places where his superiority was decisive.

The way to deter aggression is for the free community to be willing and able to respond vigorously at places and with means of its own choosing.

---

49. Herman Kahn, *On Thermonuclear War*, 1960

Some decision makers who accept the Finite Deterrence view are willing to pay for insurance against unreliability for more than political or psychological reasons. Even those who hold that war means mutual annihilation are sometimes willing for us to act beyond their beliefs—or fears.

This essay appeared in Herman Kahn, *On Thermonuclear War* (Princeton: Princeton University Press, 1961), pp. 18–23.

While this is inconsistent, it is not necessarily irrational. They understand that paper calculations can be wrong and are willing to hedge against this possibility. Sometimes these decision makers are making a distinction that (rather surprisingly) is not usually made. They may distinguish, for example, between 100 million dead and 50 million dead, and argue that the latter state is better than the former. They may distinguish between war damage which sets the economy of a country back fifty years or only ten years. *Actually when one examines the possible effects of thermonuclear war carefully, one notices that there are indeed many postwar states that should be distinguished.* If most people do not or cannot distinguish among these states it is because the gradations occur as a result of a totally bizarre circumstance—a thermonuclear war. The mind recoils from thinking hard about that; one prefers to believe it will never happen. If asked, "How does a country look on the day of the war?" the only answer a reasonable person can give is "awful." It takes an act of iron will or an unpleasant degree of detachment or callousness to go about the task of distinguishing among the possible degrees of awfulness.

But surely one can ask a more specific question. For example, "How does a country look five or ten years after the close of war, as a function of three variables: (1) the preparations made before the war, (2) the way the war started, and (3) the course of military events?" Both very sensitive and very callous individuals should be able to distinguish (and choose, perhaps) between a country which survives a war with, say 150 million people and a gross national product (GNP) of $300 billion a year, and a nation which emerges with only 50 million people and a GNP of $10 billion. The former would be the richest and the fourth largest nation in the world, and one which would be able to restore a reasonable facsimile of the prewar society; the latter would be a pitiful remnant that would contain few traces of the prewar way of life. When one asks this kind of question and examines the circumstances and possible outcomes of a future war in some detail, it appears that it is useful and necessary to make many distinctions among the results of thermonuclear war. The figures in Table 3 illustrate some simple distinctions which one may wish to make at the outset of his deliberations in this field.

Here I have tried to make the point that if we have a posture which might result in 40 million dead in a general war, and as a result of poor planning, apathy, or other causes, our posture deteriorates and a war occurs with 80 million dead, we have suffered an additional disaster, and *unnecessary* additional disaster that is almost as bad as the original disaster. If on the contrary, by spending a few billion dollars, or by being more competent or lucky, we can cut the number of dead from 40 million to 20

Table 3

Tragic but Distinguishable Postwar States

Dead	Economic Recuperation
2,000,000	1 year
5,000,000	2 years
10,000,000	5 years
20,000,000	10 years
40,000,000	20 years
80,000,000	50 years
160,000,000	100 years

Will the survivors envy the dead?

million, we have done something vastly worth doing! The survivors will not dance in the streets or congratulate each other if there have been 20 million men, women, and children killed; yet it would have been a worthwhile achievement to limit casualties to this number. It is very difficult to get this point across to laymen or experts with enough intensity to move them to action. The average citizen has a dour attitude toward planners who say that if we do thus and so it will not be 40 million dead—it will be 20 million dead. Somehow the impression is left that the planner said that there will be *only* 20 million dead. To him is often attributed the idea that this will be a tolerable or even astonishingly enough, a desirable state!

The rate of economic recuperation, like the number of lives saved, is also of extreme importance. Very few Americans can get interested in spending money or energy on preparations which, even if they worked, would result in preindustrial living standards for the survivors of a war. As will be explained later, our analysis indicates that if a country is moderately well prepared to use the assets which survive there is unlikely to be a critical level of damage to production. A properly prepared country is not "killed" by the destruction of even a major fraction of its wealth; it is more likely to be set back a given number of years in its economic growth. While recuperation times may range all the way from one to a hundred years, even the latter is far different from the "end of history."

Perhaps the most important item on the table of distinguishable states is not the numbers of dead or the number of years it takes for economic recuperation; rather, it is the question at the bottom: "Will the survivors envy the dead?" It is in some sense true that one may never recuperate from a thermonuclear war. The world may be permanently (i.e., for perhaps 10,000 years) more hostile to human life as a result of such a war. Therefore, if the question, "Can we restore the prewar conditions of life?" is asked, the answer must be "No!" But there are other relevant questions to be asked. For example: "How much more hostile will the environment be? Will it be so hostile that we or our descendants would prefer being dead than alive?" Perhaps even more pertinent is this question, "How happy or normal a life can the survivors and their descendants hope to have?" *Despite a widespread belief to the contrary, objective studies indicate that even though the amount of human tragedy would be greatly increased in the postwar world, the increase would not preclude normal and happy lives for the majority of survivors and their descendants.*

My colleagues and I came to this conclusion reluctantly; not because we did not *want* to believe it, but because it is so *hard* to believe. Thermonuclear bombs are so destructive, and destructive in so many ways, that it is difficult to imagine that there would be anything left after their large-scale use. One of my tasks with The RAND Corporation was to serve as a project leader for a study of the possibilities for alleviating the consequences of a thermonuclear war. That study was made as quantitatively and objectively as we could make it with the resources, information, and intellectual tools available to us. *We concluded that for at least the next decade or so, any picture of total world annihilation appears to be wrong, irrespective of the military course of events.*[1] Equally important, the picture of total disaster is likely to be wrong even for the two antagonists. Barring an extraordinary course for the war, or that most of the technical uncertainties turn out to lie at the disastrous end of the spectrum, one and maybe both of the antagonists should be able to restore a reasonable semblance of prewar conditions quite rapidly. Typical estimates run between one and ten years for a reasonably successful and well-prepared attacker and somewhat longer for the defender, depending mainly on the tactics of the attacker and the preparations of the defender. In the RAND study we tried to avoid using optimistic assumptions. With the exceptions to be noted, we used what were in our judgment the best values available, or we used slightly pessimistic ones. We believe that the situation is likely to be better than we indicate, rather than worse, though the latter possibility cannot be ruled out.

Exactly what is it that one must believe if he is to be convinced that it is worthwhile to buy Counterforce as Insurance? Listed below are eight phases of a thermonuclear war. If our decision makers are to justify the

1. Report on a Study of Non-Military Defense, the RAND Corporation, Report R-322-RC, July 1, 1958.

Table 4
A Complete Description of a Thermonuclear War

Includes the Analysis of:

1. Various time-phased programs for deterrence and defense and their possible impact on us, our allies, and others.
2. Wartime performance with different preattack and attack conditions.
3. Acute fallout problems.
4. Survival and patch-up.
5. Maintenance of economic momentum.
6. Long-term recuperation.
7. Postwar medical problems.
8. Genetic problems.

expense (and possible risk of strategic destabilization) that would be incurred in trying to acquire a capability for alleviating the consequences of a war, they must believe they can successfully negotiate each and every one of these phases, or that there is a reasonable chance that they can negotiate each of these phases.

I repeat: To survive a war it is necessary to negotiate *all eight* stages. If there is a catastrophic failure in any one of them, there will be little value in being able to cope with the other seven. Differences among exponents of the different strategic views can often be traced to the different estimates they make on the difficulty of negotiating one or more of these eight stages. While all of them present difficulties, most civilian military experts seem to consider the *last six* the critical ones. Nevertheless, most discussions among "classical" military experts concentrate on the *first two*. To get a sober and balanced view of the problem, one must examine all *eight*.

---

50. Robert McNamara and Counterforce "No Cities" Doctrine, 1962

A central military issue facing NATO today is the role of nuclear strategy. Four facts seem to us to dominate consideration of that role. All of

This essay appeared in Robert McNamara, "Defense Arrangements of the North Atlantic Community," U.S. *State Department Bulletin* 47 (July 9, 1962), pp. 64–70.

them point in the direction of increased integration to achieve our common defense. First, the alliance has overall nuclear strength adequate to any challenge confronting it. Second, this strength not only minimizes the likelihood of major nuclear war but makes possible a strategy designed to preserve the fabric of our societies if war should occur. Third, damage to the civil societies of the alliance resulting from nuclear warfare could be very grave. Fourth, improved non-nulcear forces, well within alliance resources, could enhance deterrence of any aggressive moves short of direct, all-out attack on Western Europe.

Let us look at the situation today. First, given the current balance of nuclear power, which we confidently expect to maintain in the years ahead, a surprise nuclear attack is simply not a rational act for any enemy. Nor would it be rational for an enemy to take the initiative in the use of nuclear weapons as an outgrowth of a limited engagement in Europe or elsewhere. I think we are entitled to conclude that either of these actions has been made highly unlikely.

Second, and equally important, the mere fact that no nation could rationally take steps leading to a nuclear war does not guarantee that a nuclear war cannot take place. Not only do nations sometimes act in ways that are hard to explain on a rational basis, but even when acting in a "rational" way they sometimes, indeed disturbingly often, act on the basis of misunderstandings of the true facts of a situation. They misjudge the way others will react and the way others will interpret what they are doing.

We must hope—indeed I think we have good reason to hope—that all sides will understand this danger and will refrain from steps that even raise the possibility of such a mutually disastrous misunderstanding. We have taken unilateral steps to reduce the likelihood of such an occurrence. We look forward to the prospect that through arms control the actual use of these terrible weapons may be completely avoided. It is a problem not just for us in the West but for all nations that are involved in this struggle we call the cold war.

For our part we feel we and our NATO allies must frame our strategy with this terrible contingency, however remote, in mind. Simply ignoring the problem is not going to make it go away.

The United States has come to the conclusion that, to the extent feasible, basic military strategy in a possible general nuclear war should be approached in much the same way that more conventional military operations have been regarded in the past. That is to say, principal military objectives, in the event of a nuclear war stemming from a major attack on the alliance, should be the destruction of the enemy's military forces, not of his civilian population.

The very strength and nature of the alliance forces make it possible for us to retain, even in the face of a massive surprise attack, sufficient reserve striking power to destroy an enemy society if driven to it. In other words, we are giving a possible opponent the strongest imaginable incentive to refrain from striking our own cities.

The strength that makes these contributions to deterrence and to the hope of deterring attack upon civil societies even in wartime does not come cheap. We are confident that our current nuclear programs are adequate and will continue to be adequate for as far into the future as we can reasonably foresee. During the coming fiscal year the United States plans to spend close to $15 billion on its nuclear weapons to assure their adequacy. For what this money buys, there is no substitute.

In particular, relatively weak national nuclear forces with enemy cities as their targets are not likely to be sufficient to perform even the function of deterrence. If they are small, and perhaps vulnerable on the ground or in the air, or inaccurate, a major antagonist can take a variety of measures to counter them. Indeed, if a major antagonist came to believe there was a substantial likelihood of its being used independently, this force would be inviting a preemptive first strike against it. In the event of war, the use of such a force against the cities of a major nuclear power would be tantamount to suicide, whereas its employment against significant military targets would have a negligible effect on the outcome of the conflict. Meanwhile the creation of a single additional national nuclear force encourages the proliferation of nuclear power with all of its attendant dangers.

In short, then, limited nuclear capabilities, operating independently, are dangerous, expensive, prone to obsolescence, and lacking in credibility as a deterrent. Clearly, the United States nuclear contribution to the alliance is neither obsolete nor dispensable.

At the same time, the general strategy I have summarized magnifies the importance of unity of planning, concentration of executive authority, and central direction. There must not be competing and conflicting strategies to meet the contingency of nuclear war. We are convinced that a general nuclear war target system is indivisible and if, despite all our efforts, nuclear war should occur, our best hope lies in conducting a centrally controlled campaign against all of the enemy's vital nuclear capabilities, while retaining reserve forces, all centrally controlled.

We know that the same forces which are targeted on ourselves are also targeted on our allies. Our own strategic retaliatory forces are prepared to respond against these forces, wherever they are and whatever their targets. This mission is assigned not only in fulfillment of our treaty commitments but also because the character of nuclear war compels it. More specifically, the United States is as much concerned with that portion of Soviet nuclear striking power that can reach Western Europe as with that portion that also can reach the United States. In short, we have undertaken the nuclear defense of NATO on a global basis. This will continue to be our objective. In the execution of this mission, the weapons in the European theater are only one resource among many.

There is, for example, the Polaris force, which we have been substantially increasing and which, because of its specially invulnerable nature, is peculiarly well suited to serve as a strategic reserve force. We have already announced the commitment of five of these ships, fully operational to the NATO Command.

This sort of commitment has a corollary for the alliance as a whole. We want and need a greater degree of alliance participation in formulating nuclear weapons policy to the greatest extent possible. We would all find it intolerable to contemplate having only a part of the strategic force launched in isolation from our main striking power.

We shall continue to maintain powerful nuclear forces for the alliance as a whole. As the President has said, "Only through such strength can we be certain of deterring a nuclear strike, or an overwhelming ground attack, upon our forces and allies."

But let us be quite clear about what we are saying and what we would have to face if the deterrent should fail. This is the almost certain prospect that, despite our nuclear strength, all of us would suffer deeply in the event of major nuclear war.

We accept our share of this responsibility within the alliance. And we believe that the combination of our nuclear strength and a strategy of controlled response gives us some hope of minimizing damage in the event that we have to fulfill our pledge. But I must point out that we do not regard this as a desirable prospect, nor do we believe that the alliance should depend solely on our nuclear power to deter actions not involving a massive commitment of any hostile force. Surely an alliance with the wealth, talent, and experience that we possess can find a better way than extreme reliance on nuclear weapons to meet our common threat. We do not believe that if the formula $E = MC^2$ had not been discovered, we should all be Communist slaves. On this question I can see no valid reason for a fundamental difference of view on the two sides of the Atlantic.

51. Presidential Directive 59, 1980

1. The so-called Countervailing Strategy, embodied as American policy on Presidential Directive 59, and signed by President Carter on 25 July 1980, has been portrayed as a significant shift in US nuclear policy away from Mutual Assured Destruction (MAD) towards a doctrine based on the capability and intent to fight—and win—a limited nuclear war fought through the medium of counterforce/point-target exchanges. In this sense, critics have pointed to a return to the "unstable" years of the McNamara Doctrine of 1962, with its implications of arms race and deterrence instability.

2. It is clear, however, that PD-59 does not in itself represent a radical shift in American targetting policy. It is, rather, the result of several years of study within the Pentagon and the National Security Council. Indeed, it is a natural offspring of studies done, and concerns evidenced, from the beginning of the first Nixon Administration. This pre-occupation with the rigidity and inadequacy of American strategic policy was expressed as early as February 1970 by President Nixon in his "State of the World" message to Congress:

Should a President, in the event of a nuclear attack, be left with the single option of ordering the mass destruction of enemy civilians in the face of the certainty that it would be followed by the mass slaughter of Americans?

3. Studies underway at that time led to the signing of National Security Decision Memorandum 242 by President Nixon on 17 January 1974. This was supplemented by the promulgation of the *Policy Guidance for the Employment of Nuclear Weapons*, signed by Secretary of Defense Schlesinger on 4 April 1974, which in turn led to a new Single Integrated Operational Plan (SIOP-5) which took effect on 1 January 1976. The general purpose behind these revisions is indicated in the *Defense Department Annual Report* for FY 1975:

What we need is a series of measured responses to aggression which bear some relationship to provocation, have prospect of terminating hostilities before general nuclear war breaks out, and leave some possibility of restoring deterrence.

This document appeared in L. Hagen, "PD-59 and the Countervailing Strategy: Continuity or Change?" Department of National Defense, Canada; Project Report No. PR 170 (Ottawa, October 1981).

4. Following the Nuclear Targetting Planning Review of 1977–79, PD-59 continues these themes, arguing for greater targetting of military assets (soft and hard), war-fighting (as opposed to recovery) industry assets, and political and C^3 centres. In this sense, there is a new emphasis on the first three general target sets of the SIOP—Soviet nuclear forces, conventional forces, military and political leadership centres at the expense, *but not to the exclusion*, of the fourth target set—the Soviet industrial and economic base. It must be recalled that fully 50 percent of the 40,000 targets in the SIOP remain dedicated to non-nuclear force targets.

5. There is little doubt, moreover, that targetting plans prior to the Nixon Administration contained plans for the targetting of Soviet military installations—although on a relatively massive and undifferentiated scale. Indeed, explicit American policy in the early McNamara period emphasized nuclear bargaining through the medium of nuclear exchanges directed at opposing military forces. This policy was de-emphasized, and that of MAD became the strategic cornerstone of American policy, when it became clear that the growth in Soviet capability, and the costs—fiscal and political—involved in maintaining a capacity for nuclear victory were unsupportable. Soviet military targets remained in the SIOP, however, with a capacity for sole attack of these within the confines of existing technology.

6. Thus, while the Countervailing Strategy *may* involve the refinement and multiplication of limited nuclear options plans using extant computer capabilities and multiple targetting memories, there is nothing *inherent* in the policy which is radically at variance with prior evolutions in American doctrine. The *tone* and *content* of the announcement may, however, have significant impact.

7. PD-59 must also be seen as part of the response of the Western Alliance to growing concern for the credibility and flexibility of the American nuclear guarantee in Central Europe. In this connection, the philosophy behind PD-59 is similar to that embodied in NATO LRTNF (Long Range Theatre Nuclear Forces) decisions and refinements in the tactical nuclear weapons area, designed to strengthen the fabric of deterrence by increasing the number and strength of ladders in the escalation process.

8. In this sense, there is general support for such a strategy within Europe, which is to be contrasted with the antagonism which greeted efforts during the McNamara period to institute somewhat similar policies at the nuclear level. This reflects increased concern in Western Europe over perceived Soviet advances at the intercontinental and Eurostrategic levels, which has partially obfuscated traditional European fears that war may be waged in Europe while preserving Soviet and American homelands.

9. The Countervailing Strategy is a reflection of the following tenets of deterrence theory:

(i)   Deterrence is best preserved by some measure of proportionality in response which maintains the credibility of the threatened use of nuclear forces by meeting each threat with roughly the same level of response. It is necessary, therefore, to develop a capability and policy which does not leave significant gaps in strategic forces or policies for their use, which can then be politically or militarily exploited by the enemy.

(ii)   Plans to use nuclear weapons in a limited fashion not only increase pre-war deterrence, but indicate a willingness and capability to respond at less than all-out levels, thereby enhancing intra-war deterrence through denying the enemy his objectives, as opposed to merely punishing him through massive societal destruction. The escalation process is thereby seen to be more rational and controllable, with the extension of some sort of strategic bargaining possible into nuclear war.

(iii)   A limited nuclear war policy enhances the credibility of *extended* deterrence in Europe and elsewhere by presenting the possibility of limited use of nuclear weapons which does not imply total destruction of the United States, and which will inflict damage against opposing military forces which will physically prevent the attainment of his military aims. In this sense, PD-59 is directed towards doctrinally minimizing the self-deterrence problems which might obtain in the event of a disaster in the theatre.

(iv)   Plans to use nuclear weapons in a limited and controlled manner do not increase the possibility of, or incentives to engage in, a nuclear war, but help prevent conflict through increased credibility, as outlined above, present the possibility—although not certainty—of military and political dialogue once conflict occurs, and by examining the question of limited use options, help ensure more responsible behaviour in a crisis situation than if no contingencies were available or considered in this area.

10. Although elements of PD-59 have characterized major concerns with, and modifications to, American nuclear doctrine since at least 1970, a series of recent developments in the strategic environment have provoked greater interest in the development of limited war strategies:

(i) improvements in Soviet conventional and Eurostrategic capabilities (such as the SS-20 and *Backfire* bomber) which simultaneously pose the perceived danger of a capability for military and political pressure in the area and heighten anxiety over the credibility of the American nuclear guarantee;

(ii) peripheral conflict involving the Soviet Union, particularly in the Persian Gulf/South-West Asia region;

(iii) growing Soviet strategic capabilities which, through a combination of numbers, accuracy, yield, and throw-weight, indicate at least a theoretical capability to destroy virtually the entire American land-based missile force by 1982–83. Particular developments of concern have been:

 a) the SS17/18/19, which are, in their warhead size and accuracy characteristics, ideal silo-busting weapons;

 b) the testing of SLBM forces in depressed-trajectory modes to reduce warning time, which can be perceived only as counterforce in their intent (primarily against SAC bases); and

 c) the capacity for cold launch—and hence re-load—on the SS18/19 systems.

11. These developments, never satisfactorily explained by the Soviet Union during the SALT process, and which generally occurred much sooner than projected, have increased American doubts about the acceptance by the USSR of the Assured Destruction model of deterrence advocated by the United States, and led to a general tendency to take prudent measures to counter the possible political or military use of Soviet strategic forces, and to, in effect, the adoption by America of a war-fightable posture as a doctrinal response.

12. PD-59, while largely a response to these developments, may also have been intended to variously:

(i) counter Republican criticism of Carter Administration defense policy;

(ii) serve as a signal to the Soviet Union that the US is willing doctrinally to meet Soviet capabilities on their own terms;

(iii) act as a general signal in the post-Afghanistan environment;

(iv) pave the way for significant upgrading of American limited nuclear option capabilities.

13. Technologies which could receive a push from PD-59, and which

would be required for any radical improvement in American capabilities to conduct limited nuclear war include:

(i)   the Mk 12A re-entry vehicles, whose deployment on Minuteman III and possibly the M-X and/or Trident D-5 missiles would be necessary to pose a significant counter-silo threat against Soviet SS18/19 ICBMs;

(ii) the Trident D-5 SLBM which could be used against hardened command and control bunkers and Soviet ICBMs such as the SS17;

(iii) M-X development in sufficient numbers to ensure survivability (PD-59 means very little without M-X deployment);

(iv) a manned penetrating bomber which could be used for follow-up strikes against point targets;

(v)   improved US command and control facilities and procedures along the lines of the Presidential Directives 53 and 58 issued in this area;

(vi) developments in ballistic missile defense; and

(vii) Increased production facilities for strategic nuclear materials and such special materials as tritium required for warhead production, which are currently in short supply; significant increases in production will be required for new weapons.

14. Only when action is taken on these various weapons programs will the real impact—intended or otherwise—of PD-59 be realized. However, it should be noted that the timing and tone of the announcement, coming during a hiatus in the detente process, and in the middle of the US Presidential election campaign, may have served to generate some undesirable side-effects. These, when coupled with certain observations on the technological level, indicate some potential problems with the Countervailing Strategy.

15. The United States does not currently have a significant capacity for major counterforce strikes against the Soviet Union. Depending on the assumptions made, and targetting scenarios adopted, a launch by the US of its entire ICBM force against Soviet ICBMs would leave 60–72% of Soviet silos and 48–57% of RVs intact.

16. If the Soviet Union ignored SALT II limits on its ICBM force and deployed new launchers at maximum current rates, and the US deployed Mk 12A warheads (in maximum numbers) on the M-X and Minuteman III forces (assuming 200 M-X and 550 Minuteman III ICBMs), a total American strike against Soviet silos in 1995 would still leave about 4200 warheads

(assuming 10 RVs per Soviet ICBM). This would represent a massive remaining capability.

17. Deployment of the Mk 12A warhead on M-X and Minuteman III ICBM (even in limited numbers: i.e., 3 warheads per 1 Minuteman III instead of its maximum capacity [7]) will allow the United States to credibly threaten remaining Soviet ICBMs following a first-strike against American ICBM forces, assuming SALT II limits. It would appear, then, that PD-59 is meant to provide doctrinal justification both for the development by the United States of a *Counter-third strike retaliatory counterforce capability* and of a first-strike *non-disarming* counterforce strike which could preserve escalation dominance and not, thereby, leave the Soviet Union in a favourable position vis à vis remaining US ICBM assets. No first-strike counterforce capability is likely to exist for the United States under any foreseeable scenario.

18. The combination of a credible Soviet counterforce capability against US ICBM fields, with a US strategic posture based on PD-59, may encourage Launch On Warning/Launch Through Attack postures in the Soviet Union. This would be highly destabilizing in crisis situations. If, however, force deployments coupled with doctrinal adjustments are made in the US, LOW/LTA incentives may be reduced on the American side.

19. It has been mentioned in public statements concerning PD-59 that Soviet command and control facilities will be targetted in the future. It is, however, difficult to see how a controlled nuclear exchange can take place if the capacity for that control is eliminated. There are, moreover, two additional problems on the Soviet side. First, political C^3 facilities are hardened and dispersed such that their location is either unknown, or such that there are so many plausible locations for actual leadership location that there would not be enough warheads to target them all. Second, such facilities as are known are co-located with major population centres; this would obviously hamper attempts at limiting the conflict.

20. There are equally serious problems with American C^3. Several points deserve emphasis here. Unless American C^3 assets are improved dramatically, the capacity for prolonged nuclear exchange is likely to be illusory. Even if improved, it is questionable whether C^3 can in any case be hardened, dispersed, or proliferated, so as to allow for the sort of intra-war control and prolonged use which more enthusiastic proponents of nuclear flexibility envisage. This relates not so much to C^3 at the systems level but at the National Command Authority-system interface. (In any case, the sort of C^3 improvements called for by a robust PD-59 will not be available until approximately 1989, if then.) The vulnerability of satellites, the "hot line," FBM SUBMARINE C^3, the NCA and airborne C^3 systems are such that it

is difficult to imagine adequate defences against electromagnetic pulse (EMP), jamming, and deception. This is not to question the need for better C^3; it is, rather, to emphasize the likely limitations which C^3 is likely to put on more ambitious strategies of flexibility and nuclear bargaining.

21. Significant casualties may be expected from any American attack against Soviet economic or military assets. For example, recent American studies demonstrate that a very limited strike against 10 Soviet oil refinery and storage centres will generate approximately 1.5 million fatalities. Many ICBM fields are co-located with large Soviet industrial conurbations and, depending on the nature of the attack and prevailing wind conditions, casualties would undoubtedly be high. Such damage levels, combined with inevitable difficulties in attack assessment might make *identification* of limited nuclear strikes difficult—perhaps provoking an all-out Soviet response—and provoke serious doubts as to how "limited" strikes of this sort would be perceived, *even assuming* accurate attack assessment.

By merely identifying the need for sufficient forces and flexibility for limited war prosecution, PD-59 does not indicate an upper limit for such efforts. As during the early McNamara years, this may be perceived as providing service rationale for increased defense spending beyond that intended, and may, under a Republican Administration, allow the movement from a counter-third strike capability towards a first-strike threat. This may be facilitated by the identification of the strategy with a previous administration; it could not then be identified as a Republican ploy.

23. An intended or unintended by-product of PD-59 may be to provide an opportunity to gauge Soviet strategic intentions. US officials are likely to watch closely to see if Soviet responses to the new strategy and associated force improvements are in the direction of:

(i)   going mobile (stabilizing if verifiable);
(ii)   going to sea (stabilizing);
(iii) deploying a new ICBM generation with increased vigour (destabilizing);
(iv) ignoring SALT II limits and building more launchers with more warheads (destabilizing); or
(v)   returning to the SALT table with an increased eagerness to negotiate limitations or reductions (stabilizing).

24. Therefore, the impact of PD-59 on SALT and deterrence stability is unclear at this point. If Soviet responses remain temperate, and US force improvements observe the limits established by current programs, there is

unlikely to be much effect. It *is* likely, in the absence of Soviet willingness
to radically decrease its ICBM capability, to increase American reluctance
to curb its own development programs. It may, however, also help to demonstrate to the USSR that doctrine and capability to the American side is
such as to *mathematically* indicate the futility of increased ICBM deployments, which may in turn ease the negotiating process. And, as a response
to the Soviet threat to US ICBMs, prudent force deployments in conjunction with the doctrinal emphasis of PD-59 may act to stabilize deterrence
by enhancing Alliance cohesion and demonstrating the will and capacity to
deny the USSR any first-strike political or military advantage. This virtue,
however, depends for its realization on the sober recognition of some of the
limitations and doubts which PD-59 evinces in the absence of concrete
force planning decisions.

52. President Ronald Reagan on Deterrence, November 23, 1982

What do we mean when we speak of nuclear deterrence? Certainly we
do not want such weapons for their own sake. We do not desire excessive
forces, or what some people have called overkill. Basically, it is a matter of
others' knowing that starting a conflict would be more costly to them than
anything they might hope to gain. And, yes, it is sadly ironic that in these
modern times it still takes weapons to prevent war. I wish it did not.

We desire peace, but peace is a goal, not a policy. Lasting peace is
what we hope for at the end of our journey; it does not describe the steps
we must take, nor the paths we should follow to reach that goal. I intend to
search for peace along two parallel paths—deterrence and arms reduction.
I believe these are the only paths that offer any real hope for an enduring
peace.

And let me say I believe that if we follow prudent policies, the risk of
nuclear conflict will be reduced. Certainly the United States will never use
its forces except in response to attack. Through the years, Soviet leaders
have also expressed a sober view of nuclear war. And if we maintain a
strong deterrent, they are extremely unlikely to launch an attack.

Now, while the policies and the capabilities have changed over the course of time, the
things we must do in order to maintain our security have not changed.

You often hear that the United States and the Soviet Union are in an

This essay appeared in the N

arms race. The truth is that while the Soviet Union has raced, we have not. In constant dollars our defense spending in the 1960's went up because of Vietnam and then it went downward through much of the 1970's. Soviet spending has gone up and up and up. In spite of a stagnating Soviet economy, Soviet leaders invest 12 to 14 percent of their country's gross national product in military spending, two to three times the level we invest.

I might add that the defense share of our United States Federal budget has gone way down, too. In 1962, when John Kennedy was President, 46 percent, almost half of the Federal budget, went to our national defense. In recent years, about one-quarter of our budget has gone to defense, while the share for social programs has nearly doubled. And most of our defense budget is spent on people, not weapons.

The combination of the Soviets' spending more and the U.S. spending proportionately less changed the military balance and weakened our deterrent. Today, in virtually every measure of military power the Soviet Union enjoys a decided advantage.

This chart (pointing to chart, "Strategic Missiles and Bombers") shows the changes in the total number of intercontinental missiles and bombers. You will see that in 1962 and in 1972, the United States forces remained about the same, even dropping some by 1982. But take a look now at the Soviet side. In 1962, at the time of the Cuban missile crisis, the Soviets could not compare with us in terms of strength. In 1972, when we signed the SALT I Treaty, we were nearly equal. But in 1982, well, that red Soviet bar stretching above the blue American bar tells the story.

I could show you chart after chart where there is a great deal of red and a much lesser amount of U.S. blue. For example, the Soviet Union has deployed a third more land-based intercontinental ballistic missiles than we have. Believe it or not, we froze our number in 1965 and have deployed no additional missiles since then.

The Soviet Union put to sea 60 new ballistic missile submarines in the last 15 years. Until last year we had not commissioned one in that same period.

The Soviet Union has built over 200 modern Backfire bombers—and is building 30 more a year. For 20 years, the United States has deployed no new strategic bombers. Many of our B-52 bombers are now older than the pilots who fly them.

The Soviet Union now has 600 of the missiles considered most threatening by both sides—the intermediate-range missiles based on land. We have none. The U.S. withdrew its intermediate-range land-based missiles from Europe almost 20 years ago.

The world has also witnessed unprecedented growth in the area of Soviet conventional forces; the Soviets far exceed us in the number of tanks, artillery pieces, aircraft and ships they produce every year. What is more, when I arrived in this office I learned that in our own forces we had planes that could not fly and ships that could not leave port, mainly for lack of spare parts and crew members.

The Soviet military buildup must not be ignored. We have recognized the problem and together with our allies we have begun to correct the imbalance. Look at this chart (accompanying chart, "Projected Defense Spending") of projected real defense spending for the next several years. Here's the Soviet line. Let us assume the Soviets' rate of spending remains at the level they have followed since the 1960's.

The blue line is the United States. If my defense proposals are passed, it will still take five years before we come close to Soviet level. Yet the modernization of our strategic and conventional forces will assure that deterrence works and peace prevails.

Our deployed nuclear forces were built before the age of microcircuits. It is not right to ask our young men and women in uniform to maintain and operate such antiques. Many have already given their lives in missile explosions and aircraft accidents caused by the old age of their equipment. We must replace and modernize our forces, and that is why I have decided to proceed with production and deployment of the new ICBM known as the MX.

Three earlier Presidents worked to develop this missile. Based on the best advice I could get, I concluded that the MX is the right missile at the right time. On the other hand, when I arrived in office, I felt the proposal on where and how to base the missile simply cost too much in terms of money, and the impact on our citizens' lives.

I have concluded, however, it is absolutely essential that we proceed to produce this missile, and that we base it in a series of closely based silos at Warren Air Force Base near Cheyenne, Wyo.

This plan requires only half as many missiles as the earlier plan and will fit in an area of only 20 square miles. It is the product of around-the-clock research that has been under way since I directed a search for a better, cheaper way. I urge the members of Congress who must pass this plan to listen and examine the facts, before they come to their own conclusion.

Some may question what modernizing our military has to do with peace. Well, as I explained earlier, a secure force keeps others from threatening us and that keeps the peace. And just as important, it also increased the products of reaching significant arms reductions with the Soviets, and that is what we really want. The United States wants deep cuts in the world's arsenal of weapons.

But unless we demonstrate the will to rebuild our strength and restore the military balance, the Soviets, since they are so far ahead, have little incentive to negotiate with us. If we had not begun to modernize, the Soviet negotiators would know we had nothing to bargain with except talk. They would know we were bluffing without a good hand because they know what cards we hold—just as we know what is in their hand.

VIII

Arms Control

*T*he term "arms control" first came into use in the 1950s in the United States among military planners and specialists interested in avoiding accidental nuclear war through miscalculation and aware of the pitfalls involved in Soviet proposals for "general and complete disarmament." Since the failure of the Baruch Plan in 1946, international control of nuclear weapons and disarmament had become utopian dreams amid the realities of a Soviet-American arms race. The devastating power of thermonuclear weapons, combined after 1957 with the rapid new delivery systems spawned by intercontinental ballistic missiles (ICBMs), made arms control imperative. Talks between Soviet and American scientists at businessman Cyrus Eaton's Nova Scotia estate, Pugwash, initiated a new tradition of bilateral talks between arms control experts on both sides of the Iron Curtain.

Neither the Soviet Union nor China accepted the concept of arms control in public, on the grounds that it was a mere bourgeois ploy to continue the arms race and that, in wartime, men and morale were of greater importance than technology. Then, in October 1962, the Soviet Union installed medium-range missiles in Cuba and precipitated a crisis that brought the powers to the brink of nuclear war. Two arms control measures that quickly resulted from the Cuban missile crisis were the partial ban on atmospheric testing of 1963 and the establishment of a "hot line" communications link between Moscow and Washington. In this way both powers sought to provide room for weapons development while avoiding misunderstandings that could lead to war by accident.

Further arms control agreements followed in the 1970s, including nonnuclear zones in Antarctica and Latin America. The key agreements were bilateral between the Soviet Union and the United States: the strategic arms limitation talks (SALT) process led, in 1972, to an agreement to limit each side's antiballistic missile sites (ABM) and to set ceilings on strategic launch vehicles. The more comprehensive SALT II agreement of 1979 was signed by Premier Brezhnev and President Carter but never ratified by the United States Senate.

Soviet-American arms control agreements helped widen the Sino-Soviet split, especially after the Soviet Union refused to provide the Chinese with a sample nuclear weapon in 1959. The Chinese were sharply critical of any "nuclear club" arrangement that would exclude the acquisition of nuclear weapons by Third World nations such as their own. In 1964 China successfully tested her first nuclear weapon and in the 1970s attained both thermonuclear and ICBM capability.

The crucial breakthroughs in arms control were technological, as well as political. New techniques of satellite photography and seismology made the violation of national security by inspection teams unnecessary and provided both sides with "national technical means of verification" for any major arms control measures.

On the 1963 test ban, see H. K. Jacobson and Eric Stein, Diplomats, Scientists, and Politicians: The United States and the Nuclear Test Ban Negotiations (Ann Arbor: University of Michigan Press, 1966). The SALT negotiations are chronicled in Thomas Wolfe, The Salt Experience (Cambridge, Mass.: Ballinger, 1979) and in Strobe Talbott, Endgame: The Inside Story of Salt II (New York: Harper & Row, 1979). A valuable personal account of the test ban negotiations is by President Eisenhower's science adviser, George B. Kistiakowsky, A Scientist at the White House (Cambridge, Mass.: Harvard University Press, 1976).

53. "Broken Arrows": Nuclear Weapons Accidents, 1950–68

Date	Weapon System	Place	Remarks
1. 5 Aug. 1950	Unspecified	Fairfield-Suison Field, California (now Travis AFB)	Unspecified.

This table appeared in *Stockholm Peace Research Institute Yearbook of World Armaments and Disarmament, 1968/1969* (New York: Humanities Press, 1969), pp. 266–68.

continued

Date	Weapon System	Place	Remarks
2. 1956	B-36 bomber	New Mexico, USA	B-36 bomber dropped an atomic bomb on barren territory near Kirtland Air Force Base, New Mexico.
3. 12 Dec. 1957	B-52 bomber	Fairchild AFB, Spokane, Washington	B-52 crashed on take off. The news report at the time specified only "a training mission," and made no mention of a weapon.
4. 5 Feb. 1958	B-47 bomber	Hunter Air Force Base, Georgia, USA	B-47 bomber, and mid-air collision, accidentally jettisoned a nuclear weapon. Weapon was in transportable condition and not capable of a nuclear explosion.
5. 12 Feb. 1958	Bomber	Off Savannah, Georgia coast, USA	Unspecified.
6. 5 March 1958	B-47 bomber	Georgia coast, USA	Bomber jettisoned an atomic bomb off the Georgia coast following a mid-air collision. This was listed as an "incident," not an "accident."
7. 11 March 1958	B-47 bomber	Florence, South Carolina	B-47 from Hunter accidentally jettisoned

continued

Date	Weapon System	Place	Remarks
			an unarmed nuclear weapon because of a malfunction of the plane's bombing system.
8. 4 Nov. 1958	B-47 bomber	Texas, USA	B-47 crashed after take-off from Dyess Air Force Base, Texas. The crash was the result of a fire.
9. 26 Nov. 1958	B-47	Louisiana, USA	B-47 caught fire and burned on the flight line at Chenrick Air Force Base, Louisiana.
10. 6 July 1959	Nuclear weapon in transit	Louisiana, USA	C-124 transport plane carrying an unarmed nuclear weapon crashed and burned on take-off from Barksdale Air Force Base in Louisiana.
11. 15 Oct. 1959	B-52 bomber	Kentucky, USA	B-52 bomber carrying two unarmed nuclear weapons collided with a KC-135 tanker plane near Glen Bean, Kentucky. The bombs were recovered undamaged.
12. 8 June 1960	Bomarc Surface-to-air missile	New Jersey, USA	Bomarc air-defense missile site at McGuire Air

continued

Date	Weapon System	Place	Remarks
			Force Base in New Jersey, caught fire. Fire and two explosions severely damaged one of the missiles, which carried a nuclear warhead.
13. 19 Jan. 1961	B-52 bomber	Monticello, Utah	B-52 exploded in the air.
14. 24 Jan. 1961	B-52 bomber	North Carolina, USA	B-52 from Seymour-Johnson Air Force Base, Goldsboro, North Carolina, carrying unarmed bombs crashed 15 miles north of the base.
15. 14 March 1961	B-52 bomber	California, USA	B-52 from Beale Air Force Base, in California, on an airborne alert training flight crashed with bombs on board.
16. 4 June 1962	Thor ICBM	Johnston Island, US Pacific Test Range	First high altitude (30 miles) thermonuclear explosion of a test series. Launch vehicle failure; ICBM's "thermonuclear device destroyed in flight." Warhead yield was one megaton.

continued

Date	Weapon System	Place	Remarks
17. 20 June 1962	Thor ICBM	Johnston Island, US Pacific Test Range	Second high altitude test shot fails. Thor missile and nuclear warhead. Again destroyed. The test was to have occurred at an altitude of 200 miles or higher. Warhead yield was again in the one megaton range. A radioactive hot spot on the fringe of the Pacific may mark for centuries the United States second failure to explode a hydrogen bomb at altitude of about 200 miles.
18. April 1963	Thresher, US nuclear powered attack submarine	US Atlantic coastline	Submarine lost; Thresher had Subroc on board which carried a nuclear warhead.
19. 13 Jan. 1964	B-52 bomber	Cumberland, Maryland USA	B-52 from Turner Air Force Base, in Georgia, crashed in Cumberland, Maryland. It carried two unarmed bombs.
20. 8 Dec. 1964	B-58 bomber	Indiana, USA	B-58 Hustler bomber caught

continued

Date	Weapon System	Place	Remarks
			fire and burned on the flight path at Bunker Hill Air Force Base, in Indiana. It carried an armed bomb.
21. 12 Oct. 1965	Nuclear weapon components	Ohio, USA	C-124 transport caught fire and burned during a refuel stop at Wright-Patterson Air Force Base, in Ohio. Nuclear weapons were not carried on the plane, but non-explosive components of nuclear systems were.
22. 17 Jan. 1966	B-52 bomber	Palomares, Spain	A B-52 and a KC-135 refueling tanker collided in midair in Palomares, Spain. B-52 crashed and 4 unarmed hydrogen bombs separated from the aircraft. One landed intact in a riverbed. The second and third bombs released radioactive material in the middle of a populated area. The fourth retrieved from the ocean April 7 after an

continued

Date	Weapon System	Place	Remarks
			intensive search. Some press reports indicated that this fourth weapon carried a 20 megaton warhead. Other reports credit all four weapons as 1.5 megatons.
23. Unspecified	Unspecified	A North African base; Morocco	Unspecified.
24. Unspecified	Unspecified	In England	Unspecified.
25. Unspecified	Unspecified	Off the US Atlantic coastline	Unspecified.
26. Unspecified	Unspecified	In the Artic	Unspecified.
27. 21 Jan. 1968	B-52 bomber	Thule, Greenland	Crash of B-52; four thermonuclear bombs lost.
28. 12 Feb. 1968	B-52 bomber	30 km north of Toronto, Canada	Alleged crash of B-52 with nuclear or thermonuclear weapons on board.

54. White House Statement on the Soviet Missile Sites in Cuba, October 26, 1962

The development of ballistic missile sites in Cuba continues at a rapid pace. Through the process of continued surveillance directed by the President, additional evidence has been acquired which clearly reflects that as of Thursday, October 25, definite buildup in these offensive missile sites continued to be made. The activity at these sites apparently is directed at achieving a full operational capability as soon as possible.

This document appeared in *Public Papers of the Presidents of the United States: John F. Kennedy, 1962* (Washington, D.C.: Government Printing Office, 1963), p. 812.

There is evidence that as of yesterday, October 25, considerable construction activity was being engaged in at the Intermediate Range Ballistic Missile sites. Bulldozers and cranes were observed as late as Thursday actively clearing new areas within the sites and improving the approach roads to the launch pads.

Since Tuesday October 23 missile related activities have continued at the Medium Range Ballistic Missile sites resulting in progressive refinements at these facilities. For example, missiles were observed parked in the open on October 23. Surveillance on October 25 revealed that some of these same missiles have now been moved from their original parked positions. Cabling can be seen running from the missile-ready tents to power generators nearby.

In summary, there is no evidence to date indicating that there is any intention to dismantle or discontinue work on these missile sites. On the contrary the Soviets are rapidly continuing their construction of missile support and launch facilities, and serious attempts are under way to camouflage their efforts.

55. White House Statement on Soviet Proposals Relating to International Security, October 27, 1962

Several inconsistent and conflicting proposals have been made by the U.S.S.R. within the last twenty-four hours, including the one just made public in Moscow. The proposal broadcast this morning involves the security of nations outside the Western Hemisphere. But it is the Western Hemisphere countries and they alone that are subject to the threat that has produced the current crisis—the action of the Soviet Government in secretly introducing offensive weapons into Cuba. Work on these offensive weapons is still proceeding at a rapid pace. The first imperative must be to deal with this immediate threat, under which no sensible negotiation can proceed.

It is therefore the position of the United States that as an urgent preliminary to consideration of any proposals work on the Cuban bases must stop; offensive weapons must be rendered inoperable; and further ship-

This document appeared in *Public Papers of the Presidents of the United States: John F. Kennedy, 1962* (Washington, D.C.: Government Printing Office, 1963), p. 813.

ment of offensive weapons to Cuba must cease—all under effective international verification.

As to proposals concerning the security of nations outside this Hemisphere, the United States and its allies have long taken the lead in seeking properly inspected arms limitation on both sides. These efforts can continue as soon as the present Soviet-created threat is ended.

NOTE: An unofficial translation of Chairman Khrushchev's message of October 27 is published in the *Department of State Bulletin* (vol. 47, p. 741). Stating that he had studied the President's reply to Mr. U. Thant "with great satisfaction," Mr. Khrushchev proposed the removal of missiles from both Cuba and Turkey under U.S. supervision. He further proposed that the Soviet Government give a solemn pledge to respect the frontiers and sovereignty of Turkey, and that the U.S. Government make a similar statement with regard to Cuba.

56. Message to Chairman Khrushchev Calling for Removal of Soviet Missiles From Cuba, October 27, 1962

Dear Mr. Chairman:

I have read your letter of October 26th with great care and welcomed the statement of your desire to seek a prompt solution to the problem. The first thing that needs to be done, however, is for work to cease on offensive missile bases in Cuba and for all weapons systems in Cuba capable of offensive use to be rendered inoperable, under effective United Nations arrangements.

Assuming this is done promptly, I have given my representatives in New York instructions that will permit them to work out this weekend—in cooperation with the Acting Secretary General and your representative—an arrangement for a permanent solution to the Cuban problem along the lines suggested in your letter of October 26th. As I read your letter, the key elements of your proposals—which seem generally acceptable as I understand them—are as follows:

1. You would agree to remove these weapons systems from Cuba under

This document appeared in *Public Papers of the Presidents of the United States: John F. Kennedy, 1962* (Washington, D.C.: Government Printing Office, 1963), pp. 813–14.

appropriate United Nations observation and supervision; and undertake, with suitable safeguards, to halt the further introduction of such weapons systems into Cuba.

2. We, on our part, would agree—upon the establishment of adequate arrangements through the United Nations to ensure the carrying out and continuation of these commitments—(a) to remove promptly the quarantine measures now in effect and (b) to give assurances against an invasion of Cuba. I am confident that other nations of the Western Hemisphere would be prepared to do likewise.

If you will give your representative similar instructions, there is no reason why we should not be able to complete these arrangements and announce them to the world within a couple of days. The effect of such a settlement on easing world tensions would enable us to work toward a more general arrangement regarding "other armaments," as proposed in your second letter which you made public. I would like to say again that the United States is very much interested in reducing tensions and halting the arms race, and if your letter signifies that you are prepared to discuss a detente affecting NATO and the Warsaw Pact, we are quite prepared to consider with our allies any useful proposals.

But the first ingredient, let me emphasize, is the cessation of work on missile sites in Cuba and measures to render such weapons inoperable, under effective international guarantees. The continuation of this threat, or a prolonging of this discussion concerning Cuba by linking these problems to the broader questions of European and world security, would surely lead to an intensification of the Cuban crisis and grave risk to the peace of the world. For this reason I hope we can quickly agree along the lines outlined in this letter and in your letter of October 26th.

John F. Kennedy

NOTE: Chairman Khrushchev's letter of October 26 was not released by the White House.

57. Message in Reply to a Broadcast by Chairman Khrushchev on the Cuban Crisis, October 28, 1962

Dear Chairman:

I am replying at once to your broadcast message of October twenty-eight, even though the official text has not yet reached me, because of the great importance I attach to moving forward promptly to the settlement of the Cuban crisis. I think that you and I, with our heavy responsibilities for the maintenance of peace, were aware that developments were approaching a point where events could have become unmanageable. So I welcome this message and consider it an important contribution to peace.

The distinguished efforts of Acting Secretary General U. Thant have greatly facilitated both our tasks. I consider my letter to you of October twenty-seventh and your reply of today as firm undertakings on the part of both our governments which should be promptly carried out. I hope that the necessary measures can at once be taken through the United Nations, as your message says, so that the United States in turn will be able to remove the quarantine measures now in effect. I have already made arrangements to report all these matters to the Organization of American States, whose members share a deep interest in a genuine peace in the Caribbean area.

You referred in your letter to a violation of your frontier by an American aircraft in the area of the Chukotskiy Peninsula. I have learned that this plane, without arms or photographic equipment, was engaged in an air sampling mission in connection with your nuclear tests. Its course was direct from Eielson Air Force Base in Alaska to the North Pole and return. In turning south, the pilot made a serious navigational error which carried him over Soviet territory. He immediately made an emergency call on open radio for navigational assistance and was guided back to his home base by the most direct route. I regret this incident and will see to it that every precaution is taken to prevent recurrence.

Mr. Chairman, both of our countries have great unfinished tasks and I know that your people as well as those of the United States can ask for nothing better than to pursue them free from the fear of war. Modern science and technology have given us the possibility of making labor fruitful beyond anything that could have been dreamed of a few decades ago.

This document appeared in *Public Papers of the Presidents of the United States: John F. Kennedy, 1962* (Washington, D.C.: Government Printing Office, 1963), pp. 814–15.

I agree with you that we must devote urgent attention to the problem of disarmament, as it relates to the whole world and also to critical areas. Perhaps now, as we step back from danger, we can together make real progress in this vital field. I think we should give priority to questions relating to the proliferation of nuclear weapons, on earth and in outer space, and to the great effort for a nuclear test ban. But we should also work hard to see if wider measures of disarmament can be agreed on and put into operation at an early date. The United States Government will be prepared to discuss these questions urgently, and in a constructive spirit, at Geneva or elsewhere.

John F. Kennedy

NOTE: An unofficial translation of Mr. Khrushchev's reply to the President's letter of October 27, is printed in the Department of State Bulletin (vol. 47, p. 743). He expressed satisfaction with the President's message of October 27 and stated his understanding of U.S. concern in connection with the weapons "you regard as offensive." The Chairman added that the Soviet Government had ordered that the weapons be dismantled, crated, and returned to Russia, and he stated that "we are prepared to reach agreement to enable U.N. representatives to verify the dismantling of these means."

Mr. Khrushchev further stated that the Soviet weapons were defensive in character and had been sent to Cuba because the Cuban people were under "continuous threat of invasion." Referring to violations of Soviet and Cuban airspace by American planes as dangerous, he concluded by affirming his belief that "reason will triumph, that war will not be unleashed, and peace and the security of the peoples will be insured."

58. Statement by the President Following the Soviet Decision to Withdraw Missiles from Cuba, October 28, 1962

I welcome Chairman Khrushchev's statesmanlike decision to stop building bases in Cuba, dismantling offensive weapons and returning them

This document appeared in *Public Papers of the Presidents of the United States: John F. Kennedy, 1962* (Washington, D.C.: Government Printing Office, 1963), p. 815.

to the Soviet Union under United Nations verification. This is an important and constructive contribution to peace.

We shall be in touch with the Secretary General of the United Nations with respect to reciprocal measures to assure peace in the Caribbean area.

It is my earnest hope that the governments of the world can, with a solution of the Cuban crisis, turn their urgent attention to the compelling necessity for ending the arms race and reducing world tensions. This applies to the military confrontation between the Warsaw Pact and NATO countries as well as to other situations in other parts of the world where tensions lead to the wasteful diversion of resources to weapons of war.

---

59. Memorandum of Understanding between the United States of America and the Union of Soviet Socialist Republics Regarding the Establishment of a Direct Communications Link, June 20, 1963 (with Annex)

Signed at Geneva June 20, 1963
Entered into force June 20, 1963

For use in time of emergency the Government of the United States of America and the Government of the Union of Soviet Socialist Republics have agreed to establish as soon as technically feasible a direct communications link between the two Governments.

Each Government shall be responsible for the arrangements for the link on its own territory. Each Government shall take the necessary steps to ensure continuous functioning of the link and prompt delivery to its head of government of any communications received by means of the link from the head of government of the other party.

Arrangements for establishing and operating the link are set forth in the Annex which is attached hereto and forms an integral part hereof.

Done in duplicate in the English and Russian languages at Geneva, Switzerland, this 20th day of June, 1963.

This document appeared in United States Arms Control and Disarmament Agency, *Arms Control and Disarmament Agreements* (Washington, D.C.: Government Printing Office, 1982), pp. 31–33.

FOR THE GOVERNMENT OF THE UNITED STATES OF AMERICA:

Acting Representative of the United States of America to the Eighteen-Nation Committee on Disarmament

FOR THE GOVERNMENT OF THE UNION OF SOVIET SOCIALIST REPUBLICS:

Acting Representative of the Union of Soviet Socialist Republics to the Eighteen-Nation Committee on Disarmament

(SEAL)

Annex

To the Memorandum of Understanding Between the United States of America and the Union of Soviet Socialist Republics Regarding the Establishment of a Direct Communications Link

The direct communications link between Washington and Moscow established in accordance with the Memorandum, and the operation of such link, shall be governed by the following provisions.

1. The direct communications link shall consist of:

a. Two terminal points with telegraph-teleprinter equipment between which communications shall be directly exchanged;

b. One full-time duplex wire telegraph circuit, routed Washington-London-Copenhagen-Stockholm-Helsinki-Moscow, which shall be used for the transmission of messages;

c. One full-time duplex radiotelegraph circuit, routed Washington-Tangier-Moscow, which shall be used for service communications and for coordination of operations between the two terminal points.

If experience in operating the direct communications link should demonstrate that the establishment of an additional wire telegraph circuit is advisable, such circuit may be established by mutual agreement between authorized representatives of both Governments.

2. In case of interruption of the wire circuit, transmission of messages shall be effected via the radio circuit, and for this purpose provision shall be made at the terminal points for the capability of prompt switching of all necessary equipment from one circuit to another.

3. The terminal points of the link shall be so equipped as to provide for the transmission and reception of messages from Moscow to Washington in the Russian language and from Washington to Moscow in the English language. In this connection, the USSR shall furnish the United States four sets of telegraph terminal equipment, including page printers, transmit-

ters, and reperforators, with one year's supply of spare parts and all necessary special tools, test equipment, operating instructions, and other technical literature, to provide for transmission and reception of messages in the Russian language.

The United States shall furnish the Soviet Union four sets of telegraph terminal equipment, including page printers, transmitters, and reperforators, with one year's supply of spare parts and all necessary special tools, test equipment, operating instructions and other technical literature, to provide for transmission and reception of messages in the English language.

Equipment described in this paragraph shall be exchanged directly between the parties without any payment being required therefor.

4. The terminal points of the direct communications link shall be provided with encoding equipment. For the terminal point in the USSR, four sets of such equipment (each capable of simplex operation), with one year's supply of spare parts, with all necessary special tools, test equipment, operating instructions and other technical literature, and with all necessary blank tape, shall be furnished by the United States to the USSR against payment of the cost thereof by the USSR.

The USSR shall provide for preparation and delivery of keying tapes to the terminal point of the link in the United States for reception of messages from the USSR. The United States shall provide for the preparation and delivery of keying tapes to the terminal point of the link in the USSR for reception of messages from the United States. Delivery of prepared keying tapes to the terminal points of the link shall be effected through the Embassy of the USSR in Washington (for the terminal of the link in the USSR) and through the Embassy of the United States in Moscow (for the terminal of the link in the United States).

5. The United States and USSR shall designate the agencies responsible for the arrangements regarding the direct communications link, for its technical maintenance, continuity and reliability, and for the timely transmission of messages.

Such agencies may, by mutual agreement, decide matters and develop instructions relating to the technical maintenance and operation of the direct communications link and effect arrangements to improve the operation of the link.

6. The technical parameters of the telegraph circuits of the link and of the terminal equipment, as well as the maintenance of such circuits and equipment shall be in accordance with CCITT and CCIR recommendations.

Transmission and reception of messages over the direct communications link shall be effected in accordance with applicable recommendations of international telegraph and radio communications regulations, as well as with mutually agreed instructions.

    7. The costs of the direct communications link shall be borne as follows:

        a. The USSR shall pay the full cost of leasing the portion of the telegraph circuit from Moscow to Helsinki and 50% of the cost of leasing the portion of the telegraph circuit from Helsinki to London. The United States shall pay the full cost of leasing the portion of the telegraph circuit from London to Helsinki.

    Payment of the cost of leasing the radio telegraph circuit between Washington and Moscow shall be effected without any transfer of payments between the parties. The USSR shall bear the expenses relating to the transmission of messages from Moscow to Washington. The United States shall bear the expenses relating to the transmission of messages from Washington to Moscow.

60. Treaty on the Nonproliferation of Nuclear Weapons, March 5, 1970

Signed at Washington, London, and Moscow July 1, 1968
Ratification advised by U.S. Senate March 13, 1969
Ratified by U.S. President November 24, 1969
U.S. ratification deposited at Washington, London,
    and Moscow March 5, 1970
Proclaimed by U.S. President March 5, 1970
Entered into force March 5, 1970

    The States concluding this Treaty, hereinafter referred to as the "Parties to the Treaty,"

    Considering the devastation that would be visited upon all mankind by a nuclear war and the consequent need to make every effort to avert the danger of such a war and to take measures to safeguard the security of peoples,

    Believing that the proliferation of nuclear weapons would seriously enhance the danger of nuclear war,

This treaty is F. R. Document 76-3382; filed February 2, 1976.

In conformity with resolutions of the United Nations General Assembly calling for the conclusion of an agreement on the prevention of wider dissemination of nuclear weapons,

Undertaking to cooperate in facilitating the application of International Atomic Energy Agency safeguards on peaceful nuclear activities,

Expressing their support for research, development and other efforts to further the application, within the framework of the International Atomic Energy Agency safeguards system, of the principle of safeguarding effectively the flow of source and special fissionable materials by the use of instruments and other techniques at certain strategic points,

Affirming the principle that the benefits of peaceful applications of nuclear technology, including any technological by-products which may be derived by nuclear-weapon States from the development of nuclear explosive devices, should be available for peaceful purposes to all Parties of the Treaty, whether nuclear-weapon or non-nuclear weapon States,

Convinced that, in furtherance of this principle, all Parties to the Treaty are entitled to participate in the fullest possible exchange of scientific information for, and to contribute alone or in cooperation with other States to, the further development of the applications of atomic energy for peaceful purposes,

Declaring their intention to achieve at the earliest possible date the cessation of the nuclear arms race and to undertake effective measures in the direction of nuclear disarmament,

Urging the cooperation of all States in the attainment of this objective,

Recalling the determination expressed by the Parties to the 1963 Treaty banning nuclear weapon tests in the atmosphere, in outer space and under water in its Preamble to seek to achieve the discontinuance of all test explosions of nuclear weapons for all time and to continue negotiations to this end,

Desiring to further the easing of international tension and the strengthening of trust between States in order to facilitate the cessation of the manufacture of nuclear weapons, the liquidation of all their existing stockpiles, and the elimination from national arsenals of nuclear weapons and the means of their delivery pursuant to a treaty on general and complete disarmament under strict and effective international control,

Recalling that, in accordance with the Charter of the United Nations States must refrain in their international relations from the threat of use of force against the territorial integrity or political independence of any State, or in any other manner inconsistent with the Purposes of the United Nations, and that the establishment and maintenance of international peace and security are to be promoted with the least diversion for armaments of the world's human and economic resources,

Have agreed as follows:

Article I

Each nuclear-weapon State Party to the Treaty undertakes not to transfer to any recipient whatsoever nuclear weapons or other nuclear explosive devices or control over such weapons or explosive devices directly, or indirectly; and not in any way to assist, encourage, or induce any non-nuclear-weapon State to manufacture or otherwise acquire nuclear weapons or other nuclear explosive devices, or control over such weapons or explosive devices.

Article II

Each non-nuclear-weapon State Party to the Treaty undertakes not to receive the transfer from any transfer or whatsoever of nuclear weapons or other nuclear explosive devices or of control over such weapons or explosive devices directly, or indirectly; not to manufacture or otherwise acquire nuclear weapons or other nuclear explosive devices; and not to seek or receive any assistance in the manufacture of nuclear weapons or other nuclear explosive devices.

Article III

1. Each non-nuclear-weapon State Party to the Treaty undertakes to accept safe-guards, as set forth in an agreement to be negotiated and concluded with the international Atomic Energy Agency in accordance with the Statute of the International Atomic Energy Agency and the Agency's safeguards system, for the exclusive purpose of verification of the fulfillment of it obligations assumed under this Treaty with a view to preventing diversion on nuclear energy from peaceful uses to nuclear weapons or other nuclear explosive devices. Procedures for the safeguards required by this article shall be followed with respect to source or fissionable material whether it is being produced, processed or used in any principle nuclear facility or is outside any such facility. The safeguards required by this article shall be applied to all source or special fissionable material in all peaceful nuclear activities within the territory of such State, under its jurisdiction, or carried out under its control anywhere.

2. Each State Party to the Treaty undertakes not to provide: (a) source or special fissionable material, or (b) equipment or material especially designed or prepared for the processing, use or production of special fissionable material, to any non-nuclear-weapon State for peaceful purposes,

unless the source or special fissionable material shall be subject to the safeguards required by this article.

3. The safeguards required by this article shall be implemented in a manner designed to comply with Article IV of this Treaty, and to avoid hampering the economic or technological development of the Parties or international cooperation in the field of peaceful nuclear activities, including the international exchange of nuclear material and equipment for the processing, use or production of nuclear material for peaceful purposes in accordance with the provisions of this article and the principle of safeguarding set forth in the Preamble of the Treaty.

4. Non-nuclear-weapon States Party to the Treaty shall conclude agreements with the International Atomic Energy Agency to meet the requirements of this article either individually or together with other States in accordance with the Statute of the International Atomic Energy Agency. Negotiation of such agreements shall commence within 180 days from the original entry into force of this Treaty. For States depositing their instruments of ratification or accession after the 180-day period, negotiation of such agreements shall commence not later than the date of such deposit. Such agreements shall enter into force not later than eighteen months after the date of initiation of negotiations.

Article IV

1. Nothing in the Treaty shall be interpreted as affecting the inalienable right of all the Parties to the Treaty to develop research, production and use of nuclear energy for peaceful purposes without discrimination and in conformity with Articles I and II of this Treaty.

2. All the Parties to the Treaty undertake to facilitate, and have the right to participate in, the fullest possible exchange of equipment, materials and scientific and technological information for the peaceful uses of nuclear energy. Parties to the Treaty in a position to do so shall also cooperate in contributing alone or together with other States or international organizations to the further development of the applications of nuclear energy for peaceful purposes, especially in the territories of non-nuclear-weapon States Party to the Treaty, with due consideration for the needs of the developing areas of the world.

Article V

Each Party to the Treaty undertakes to take appropriate measures to ensure that, in accordance with this Treaty, under appropriate international observation and through appropriate international procedures, potential benefits from any peaceful applications of nuclear explosions will be made available to non-nuclear-weapon States Party to the Treaty on a nondiscriminatory basis and that the charge to such Parties for the explosive devices used will be as low as possible and exclude any charge for research and development. Non-nuclear-weapon States Party to the Treaty shall be able to obtain such benefits, pursuant to a special international agreement or agreements, through an appropriate international body with adequate representation of non-nuclear-weapon States. Negotiations on this subject shall commence as soon as possible after the Treaty enters into force. Non-nuclear-weapon States Party to the Treaty so desiring may also obtain such benefits pursuant to bilateral agreements.

Article VI

Each of the Parties to the Treaty undertakes to pursue negotiations in good faith on effective measures relating to cessation of the nuclear arms race at an early date and to nuclear disarmament, and on a treaty on general and complete disarmament under strict and effective international control.

Article VII

Nothing in this Treaty affects the right of any group of States to conclude regional treaties in order to assure the total absence of nuclear weapons in their respective territories.

Article VIII

1. Any Party to the Treaty may propose amendments to this Treaty. The text of any proposed amendment shall be submitted to the Depositary Governments which shall circulate it to all Parties to the Treaty. Thereupon, if requested to do so by one-third or more of the Parties to the Treaty, the Depositary Governments shall convene a conference to which they shall invite all the Parties to the Treaty, to consider such an amendment.

2. Any amendment to this Treaty must be approved by a majority of the votes of all the Parties to the Treaty, including the votes of all nuclear-weapon States Party to the Treaty and all other Parties which, on the date the amendment is circulated, are members of the Board of Governors of the International Atomic Energy Agency. The amendment shall enter into force for each Party that deposits its instrument of ratification of the amendment upon the deposit of such instruments of ratification by a major-

ity of all the Parties, including the instruments of ratification of all nuclear-weapon States Party to the Treaty and all other Parties which, on the date the amendment is circulated, are members of the Board of Governors of the International Atomic Energy Agency. Thereafter, it shall enter into force for any other Party upon the deposit of its instrument of ratification of the amendment.

3. Five years after the entry into force of this Treaty, a conference of Parties to the Treaty shall be held in Geneva, Switzerland, in order to review the operation of this Treaty with a view to assuring that the purposes of the Preamble and the provisions of the Treaty are being realized. At intervals of five years thereafter, a majority of the Parties to the Treaty may obtain, by submitting a proposal to this effect to the Depositary Governments, the convening of further conferences with the same objective of reviewing the operation of the Treaty.

Article IX

1. This Treaty shall be open to all States for signature. Any State which does not sign the Treaty before its entry into force in accordance with paragraph 3 of this Article may accede to it at any time.

2. This Treaty shall be subject to ratification by signatory States. Instruments of ratification and instruments of accession shall be deposited with the Governments of the United States of America, the United Kingdom of Great Britain and Northern Ireland and the Union of Soviet Socialist Republics, which are hereby designated the Depositary Governments.

3. This Treaty shall enter into force after its ratification by the States, the Governments of which are designated Depositaries of the Treaty, and forty other States signatory to this Treaty and the deposit of their instruments of ratification. For the purposes of this Treaty, a nuclear-weapon State is one which has manufactured and exploded a nuclear weapon or other nuclear explosive device prior to January 1, 1967.

4. For States whose instruments of ratification or accession are deposited subsequent to the entry into force of this Treaty, it shall enter into force on the date of the deposit of their instruments of ratification or accession.

5. The Depositary Governments shall promptly inform all signatory and acceding States of the date of each signature, the date of deposit of each instrument of ratification or of accession, the date of the entry into force of this Treaty, and the date of receipt of any requests for convening a conference or other notices.

6. This Treaty shall be registered by the Depositary Governments pursuant to article 102 of the Charter of the United Nations.

Article X

1. Each Party shall in exercising its national sovereignty have the right to withdraw from the Treaty if it decides that extraordinary events, related to the subject matter of this Treaty, have jeopardized the supreme interests of its country. It shall give notice of such withdrawals of all other Parties to the Treaty and to the United Nations Security Council three months in advance. Such notice shall include a statement of the extraordinary events it regards as having jeopardized its supreme interests.

2. Twenty-five years after the entry into force of the Treaty, a conference shall be convened to decide whether the Treaty shall continue in force indefinitely, or shall be extended for an additional fixed period or periods. The decision shall be taken by a majority of the Parties to the Treaty.

Article XI

This Treaty, the English, Russian, French, Spanish and Chinese texts of which are equally authentic, shall be deposited in the archives of the Depositary Governments. Duly certified copies of this Treaty shall be transmitted by the Depositary Governments to the Governments of the signatory and acceding States.

61. Treaty between the United States of America and the Union of Soviet Socialist Republics on the Limitation of Anti-Ballistic Missile Systems, Moscow, May 26, 1972

The United States of America and the Union of Soviet Socialist Republics, hereinafter referred to as the Parties,

Proceeding from the premise that nuclear war would have devastating consequences for all mankind,

Considering that effective measures to limit anti-ballistic missile systems would be a substantial factor in curbing the race in strategic offensive arms and would lead to a decrease in the risk of outbreak of war involving nuclear weapons,

Proceeding from the premise that the limitation of anti-ballistic missile

This document appeared in United States Arms Control and Disarmament Agency, *Arms Control and Disarmament Agreements* (Washington, D.C.: Government Printing Office, 1982), pp. 139–42.

systems, as well as certain agreed measures with respect to the limitation of strategic offensive arms,[1] would contribute to the creation of more favorable conditions for further negotiations on limiting strategic arms,

Mindful of their obligations under Article VI of the Treaty on the Non-Proliferation of Nuclear Weapons,[2]

Declaring their intention to achieve at the earliest possible date the cessation of the nuclear arms race and to take effective measures toward reductions in strategic arms, nuclear disarmament, and general and complete disarmament,

Desiring to contribute to the relaxation of international tension and the strengthening of trust between States,

Have agreed as follows:

Article I

1. Each Party undertakes to limit anti-ballistic missile (ABM) systems and to adopt other measures in accordance with the provisions of this Treaty.

2. Each Party undertakes not to deploy ABM systems for a defense of the territory of its country and not to provide a base for such a defense, and not to deploy ABM systems for defense of an individual region except as provided for in Article III of this Treaty.

Article II

1. For the purposes of this Treaty an ABM system is a system to counter strategic ballistic missiles or their elements in flight trajectory, currently consisting of:

(a) ABM interceptor missiles, which are interceptor missiles constructed and deployed for an ABM role, or of a type tested in an ABM mode;

(b) ABM launchers, which are launchers constructed and deployed for launching ABM interceptor missiles; and

(c) ABM radars, which are radars constructed and deployed for an ABM role, or of a type tested in an ABM mode.

2. The ABM system components listed in paragraph 1 of this Article include those which are:

(a) operational;

(b) under construction;

---

1. For interim agreement and protocol between the United States and the Soviet Union, signed May 26, 1972, see TIAS 7504; 23 UST.

2. TIAS 6839; 21 UST 490.

(c) undergoing testing;

(d) undergoing overhaul, repair or conversion; or

(e) mothballed.

Article III

Each Party undertakes not to deploy ABM systems or their components except that:

(a) within one ABM system deployment area having a radius of one hundred and fifty kilometers and centered on the Party's national capital, a Party may deploy: (1) no more than one hundred ABM launchers and no more than one hundred ABM interceptor missiles at launch sites, and (2) ABM radars within no more than six ABM radar complexes, the area of each complex being circular and having a diameter of no more than three kilometers; and

(b) within one ABM system deployment area having a radius of one hundred and fifty kilometers and containing ICBM silo launchers, a Party may deploy: (1) no more than one hundred ABM launchers and no more than one hundred ABM interceptor missiles at launch sites, (2) two large phased-array ABM radars comparable in potential to corresponding ABM radars operational or under construction on the date of signature of the Treaty in an ABM system deployment area containing ICBM silo launchers, and (3) no more than eighteen ABM radars each having a potential less than the potential of the smaller of the above-mentioned two large phased-array ABM radars.

Article IV

The limitations provided for in Article III shall not apply to ABM systems or their components used for development or testing, and located within current or additionally agreed test ranges. Each Party may have no more than a total of fifteen ABM launchers at test ranges.

Article V

1. Each Party undertakes not to develop, test, or deploy ABM systems or components which are sea-based, air-based, space-based, or mobile land-based.

2. Each Party undertakes not to develop, test, or deploy ABM launchers for launching more than one ABM interceptor missile at a time from each launcher, nor to modify deployed launchers to provide them with such a capability, nor to develop, test, or deploy automatic or semi-automatic or other similar systems for rapid reload of ABM launchers.

Article VI

To enhance assurance of the effectiveness of the limitations on ABM systems and their components provided by this Treaty, each Party undertakes:

(a) not to give missiles, launchers, or radars, other than ABM interceptor missiles, ABM launchers, or ABM radars, capabilities to counter strategic ballistic missiles or their elements in flight trajectory, and not to test them in an ABM mode; and

(b) not to deploy in the future radars for early warning of strategic ballistic missiles attack except at locations along the periphery of its national territory and oriented outward.

Article VII

Subject to the provisions of this Treaty, modernization and replacement of ABM systems or their components may be carried out.

Article VIII

ABM systems or their components in excess of the numbers or outside the areas specified in this Treaty, as well as ABM systems or their components prohibited by this Treaty, shall be destroyed or dismantled under agreed procedures within the shortest possible agreed period of time.

Article IX

To assure the viability and effectiveness of this Treaty, each Party undertakes not to transfer to other States, and not to deploy outside its national territory, ABM systems or their components limited by this Treaty.

Article X

Each Party undertakes not to assume any international obligations which would conflict with this Treaty.

Article XI

The Parties undertake to continue active negotiations for limitations on strategic offensive arms.

Article XII

1. For the purpose of providing assurance of compliance with the provisions of this Treaty, each Party shall use national technical means of veri-

fication at its disposal in a manner consistent with generally recognized principles of international law.

2. Each Party undertakes not to interfere with the national technical means of verification of the other Party operating in accordance with paragraph 1 of this Article.

3. Each Party undertakes not to use deliberate concealment measures which impede verification by national technical means of compliance with the provisions of this Treaty. This obligation shall not require changes in current construction, assembly, conversion, or overhaul practices.

Article XIII

1. To promote the objectives and implementation of the provisions of this Treaty, the Parties shall establish promptly a Standing Consultative Commission, within the framework of which they will:

(a) consider questions concerning compliance with the obligations assumed and related situations which may be considered ambiguous;

(b) provide on a voluntary basis such information as either Party considers necessary to assure confidence in compliance with the obligations assumed;

(c) consider questions involving unintended interference with national technical means of verification;

(d) consider possible changes in the strategic situation which have a bearing on the provisions of this Treaty;

(e) agree upon procedures and dates for destruction or dismantling of ABM systems or their components in cases provided for by the provisions of this Treaty;

(f) consider, as appropriate, possible proposals for further increasing the viability of this Treaty, including proposals for amendments in accordance with the provisions of this Treaty;

(g) consider, as appropriate, proposals for further measures aimed at limiting strategic arms.

2. The Parties through consultation shall establish, and may amend as appropriate, Regulations for the Standing Consultative Commission governing procedures, composition and other relevant matters.

Article XIV

1. Each Party may propose amendments to this Treaty. Agreed amendments shall enter into force in accordance with the procedures governing the entry into force of this Treaty.

2. Five years after entry into force of this Treaty, and at five year intervals thereafter, the Parties shall together conduct a review of this Treaty.

Article XV

1. This Treaty shall be of unlimited duration.

2. Each Party shall, in exercising its national sovereignty, have the right to withdraw from this Treaty if it decides that extraordinary events related to the subject matter of this Treaty have jeopardized its supreme interests. It shall give notice of its decision to the other Party six months prior to withdrawal from the Treaty. Such notice shall include a statement of the extraordinary events the notifying Party regards as having jeopardized its supreme interests.

Article XVI

1. This Treaty shall be subject to ratification in accordance with the constitutional procedures of each Party. The Treaty shall enter into force on the day of the exchange of instruments of ratification.

2. This Treaty shall be registered pursuant to Article 102 of the Charter of the United Nations.[3]

Done at Moscow on May 26, 1972, in two copies, each in the English and Russian languages, both texts being equally authentic.

FOR THE UNITED STATES OF AMERICA:[4]

President of the United States of America

FOR THE UNION OF SOVIET SOCIALIST REPUBLICS:[5]

General Secretary of the Central Committee of the CPSU

3. TS 993; 59 Stat. 1052.
4. Richard Nixon.
5. L. I. Brezhnev.

---

62. Interim Agreement between the United States of America and the Union of Soviet Socialist Republics on Certain Measures with Respect to the Limitation of Strategic Offensive Arms, Moscow, May 26, 1972 (and Protocol)

The United States of America and the Union of Soviet Socialist Republics, hereinafter referred to as the Parties,

This document appeared in United States Arms Control and Disarmament Agency, *Arms Control and Disarmament Agreements* (Washington, D.C.: Government Printing Office, 1982), pp. 150–53.

Convinced that the Treaty on the Limitation of Anti-Ballistic Missile Systems[1] and this Interim Agreement on Certain Measures with Respect to the Limitation of Strategic Offensive Arms will contribute to the creation of more favorable conditions for active negotiations on limiting strategic arms as well as to the relaxation of international tension and the strengthening of trust between States,

Taking into account the relationship between strategic offensive and defensive arms,

Mindful of their obligations under Article VI of the Treaty on the Non-Proliferation of Nuclear Weapons,[2]

Have agreed as follows:

Article I

The Parties undertake not to start construction of additional fixed land-based intercontinental ballistic missile (ICBM) launchers after July 1, 1972.

Article II

The Parties undertake not to convert land-based launchers for light ICBM's, or for ICBM's of older types deployed prior to 1964, into land-based launchers for heavy ICBM's of types deployed after that time.

Article III

The Parties undertake to limit submarine-launched ballistic missile (SLBM) launchers and modern ballistic missile submarines to the numbers operational and under construction on the date of signature of this Interim Agreement, and in addition to launchers and submarines constructed under procedures established by the Parties as replacements for an equal number of ICBM launchers of older types deployed prior to 1964 or for launchers on older submarines.

Article IV

Subject to the provisions of this Interim Agreement, modernization and replacement of strategic offensive ballistic missiles and launchers covered by this Interim Agreement may be undertaken.

1. TIAS 7503; 23 UST.
2. TIAS 6839; 21 UST 490.

Article V

1. For the purpose of providing assurance of compliance with the provisions of this Interim Agreement, each Party shall use national technical means of verification at its disposal in a manner consistent with generally recognized principles of international law.

2. Each Party undertakes not to interfere with the national technical means of verification of the other Party operating in accordance with paragraph 1 of this Article.

3. Each Party undertakes not to use deliberate concealment measures which impede verification by national technical means of compliance with the provisions of this Interim Agreement. This obligation shall not require changes in current construction, assembly, conversion, or overhaul practices.

Article VI

To promote the objectives and implementation of the provisions of this Interim Agreement, the Parties shall use the Standing Consultative Commission established under Article XIII of the Treaty on the Limitation of Anti-Ballistic Missile Systems in accordance with the provisions of that Article.

Article VII

The Parties undertake to continue active negotiations for limitations on strategic offensive arms. The obligations provided for in this Interim Agreement shall not prejudice the scope or terms of the limitations on strategic offensive arms which may be worked out in the course of further negotiations.

Article VIII

1. This Interim Agreement shall enter into force[3] upon exchange of written notices of acceptance by each Party, which exchange shall take place simultaneously with the exchange of instruments of ratification of the Treaty on the Limitation of Anti-Ballistic Missile Systems.

2. This Interim Agreement shall remain in force for a period of five years unless replaced earlier by an agreement on more complete measures limiting strategic offensive arms. It is the objective of the Parties to conduct active follow-on negotiations with the aim of concluding such an agreement as soon as possible.

3. October 3, 1972.

3. Each Party shall, in exercising its national sovereignty, have the right to withdraw from this Interim Agreement if it decides that extraordinary events related to the subject matter of this Interim Agreement have jeopardized its supreme interests. It shall give notice of its decision to the other Party six months prior to withdrawal from this Interim Agreement. Such notice shall include a statement of the extraordinary events the notifying Party regards as having jeopardized its supreme interests.

Done at Moscow on May 26, 1972, in two copies, each in the English and Russian languages, both texts being equally authentic.

FOR THE UNITED STATES OF AMERICA:[4]

President of the United States of America

FOR THE UNION OF SOVIET SOCIALIST REPUBLICS:[5]

General Secretary of the Central Committee of the CPSU

Protocol

To the Interim Agreement between
the United States of America and the Union of Soviet
Socialist Republics on Certain Measures with Respect
to the Limitation of Strategic Offensive Arms

The United States of America and the Union of Soviet Socialist Republics, hereinafter referred to as the Parties,

Having agreed on certain limitations relating to submarine-launched ballistic missile launchers and modern ballistic missile submarines, and to replacement procedures, in the Interim Agreement,

Have agreed as follows:

The Parties understand that, under Article III of the Interim Agreement, for the period during which that Agreement remains in force:

The US may have no more than 710 ballistic missile launchers on submarines (SLBM's) and no more than 44 modern ballistic missile submarines. The Soviet Union may have no more than 950 ballistic missile launchers on submarines and no more than 62 modern ballistic missile submarines.

Additional ballistic missile launchers on submarines up to the above-mentioned levels, in the U.S.—over 656 ballistic missile launchers on nuclear-powered submarines, and in the U.S.S.R.—over 740 ballistic missile launchers on nuclear-powered submarines, operational and under con-

4. Richard Nixon.
5. L. I. Brezhnev.

struction, may become operational as replacements for equal numbers of ballistic missile launchers of older types deployed prior to 1964 or of ballistic missile launchers on older submarines.

The deployment of modern SLBM's on any submarine, regardless of type, will be counted against the total level of SLBM's permitted for the U.S. and the U.S.S.R.

This Protocol shall be considered an integral part of the Interim Agreement.

Done at Moscow this 26th day of May, 1972.

FOR THE UNITED STATES
OF AMERICA:

President of the United States
of America

FOR THE UNION OF SOVIET
SOCIALIST REPUBLICS

General Secretary of the
Central Committee of the CPSU

63. USA-USSR Agreement to Reduce the Risk of Nuclear War Outbreak, September 30, 1972

The United States of America and the Union of Soviet Socialist Republics, hereinafter referred to as the Parties:

Taking into account the devastating consequences that nuclear war would have for all mankind, and recognizing the need to exert every effort to avert the risk of outbreak of such a war, including measures to guard against accidental or unauthorized use of nuclear weapons,

Believing that agreement on measures for reducing the risk of outbreak of nuclear war serves the interests of strengthening international peace and security, and is in no way contrary to the interests of any other country,

Bearing in mind that continued efforts are also needed in the future to seek ways of reducing the risk of outbreak of nuclear war,

Have agreed as follows:

Article 1

Each Party undertakes to maintain and to improve, as it deems necessary, its existing organizational and technical arrangements to guard against the accidental or unauthorized use of nuclear weapons under its control.

This document appeared in United States Arms Control and Disarmament Agency, *Arms Control and Disarmament Agreements* (Washington, D.C.: Government Printing Office, 1982) pp. 111–12.

Article 2

The Parties undertake to notify each other immediately in the event of an accidental, unauthorized or any other unexplained incident involving a possible detonation of a nuclear weapon which could create a risk of outbreak of nuclear war. In the event of such an incident, the Party whose nuclear weapon is involved will immediately make every effort to take necessary measures to render harmless or destroy such weapon without its causing damage.

Article 3

The Parties undertake to notify each other immediately in the event of detection by missile warning systems of unidentified objects, or in the event of signs of interference with these systems or with related communications facilities, if such occurrences could create a risk of outbreak of nuclear war between the two countries.

Article 4

Each Party undertakes to notify the other Party in advance of any planned missile launches if such launches will extend beyond its national territory in the direction of the other Party.

Article 5

Each Party, in other situations involving unexplained nuclear incidents, undertakes to act in such a manner as to reduce the possibility of its actions being misinterpreted by the other Party. In any such situation, each Party may inform the other Party or request information when, in its view, this is warranted by the interests of averting the risk of outbreak of nuclear war.

Article 6

For transmission of urgent information, notifications and requests for information in situations requiring prompt clarification, the Parties shall make primary use of the Direct Communications Link between the Governments of the United States of America and the Union of Soviet Socialist Republics.[1]

For transmission of other information, notifications and requests for information, the Parties, at their own discretion, may use any communica-

1. See TIAS 5362, 14 UST 825; 22 UST.

tions facilities, including diplomatic channels, depending on the degree of urgency.

Article 7

The Parties undertake to hold consultations, as mutually agreed, to consider questions relating to implementation of the provisions of this Agreement, as well as to discuss possible amendments thereto aimed at further implementation of the purposes of this Agreement.

Article 8

This Agreement shall be of unlimited duration.

Article 9

This Agreement shall enter into force upon signature.

Done at Washington on September 30, 1971, in two copies, each in the English and Russian languages, both texts being equally authentic.

FOR THE UNITED STATES
OF AMERICA:[2]

FOR THE UNION OF SOVIET
SOCIALIST REPUBLICS:[3]

2. William P. Rogers.
3. A. Gromyko.

---

64. Protocol to the Treaty between the United States of America and the Union of Soviet Socialist Republics on the Limitation of Anti-Ballistic Missile Systems, Moscow, July 3, 1974

The United States of America and the Union of Soviet Socialist Republics, hereinafter referred to as the Parties,

Proceeding from the Basic Principles of Relations between the United States of America and the Union of Soviet Socialist Republics signed on May 29, 1972,

This document appeared in United States Arms Control and Disarmament Agency, *Arms Control and Disarmament Agreements* (Washington, D.C.: Government Printing Office, 1982), pp. 162–63.

Desiring to further the objectives of the Treaty between the United States of America and the Union of Soviet Socialist Republics on the Limitation of Anti-Ballistic Missile Systems signed on May 26, 1972, hereinafter referred to as the Treaty.

Reaffirming their conviction that the adoption of further measures for the limitation of strategic arms would contribute to strengthening international peace and security.

Proceeding from the premise that further limitation of anti-ballistic missile systems will create more favorable conditions for the completion of work on a permanent agreement on more complete measures for the limitation of strategic offensive arms,

Have agreed as follows:

Article I

1. Each Party shall be limited at any one time to a single area out of the two provided in Article III of the Treaty for deployment of anti-ballistic missile (ABM) systems or their components and accordingly shall not exercise its right to deploy an ABM system or its components in the second of the two ABM system deployment areas permitted by Article III of the Treaty, except as an exchange of one permitted area for the other in accordance with Article II of this Protocol.

2. Accordingly, except as permitted by Article II of this Protocol: The United States of America shall not deploy an ABM system or its components in the area centered on its capital, as permitted by Article III(s) of the Treaty, and the Soviet Union shall not deploy an ABM system or its components in the deployment area of intercontinental ballistic missile (ICBM) silo launchers permitted by Article III of the Treaty.

Article II

1. Each Party shall have the right to dismantle or destroy its ABM system and the components thereof in the area where they are presently deployed and to deploy an ABM system or its components in the alternative area permitted by Article III of the Treaty provided that prior to initiation of construction, notification is given in accord with the procedure agreed to by the Standing Consultative Commission, during the year beginning October 3, 1977, and ending October 2, 1978, or during any year which commences at five year intervals thereafter, those being the years for periodic review of the Treaty, as provided in Article XIV of the Treaty. This right may be exercised only once.

2. Accordingly, in the event of such notice, the United States would

have the right to dismantle or destroy the ABM system and its components in the deployment area of ICBM silo launchers and to deploy an ABM system or its components in an area centered on its capital, as permitted by Article III(a) of the Treaty, and the Soviet Union would have the right to dismantle or destroy the ABM system and its components in the area centered on its capital and to deploy an ABM system or its components in an area containing ICBM silo launchers, as permitted by Article III(b) of the Treaty.

3. Dismantling or destruction and deployment of ABM systems or their components and the notification thereof shall be carried out in accordance with Article VIII of the ABM Treaty and procedures agreed to in the Standing Consultative Commission.

Article III

The rights and obligations established by the Treaty remain in force and shall be complied with by the Parties except to the extent modified by this Protocol. In particular, the deployment of an ABM system or its components within the area selected shall remain limited by the levels and other requirements established by the Treaty.

Article IV

This Protocol shall be subject to ratification in accordance with the constitutional procedures of each Party. It shall enter into force on the day of the exchange of instruments of ratification and shall thereafter be considered an integral part of the Treaty.

Done at Moscow on July 3, 1974, in duplicate, in the English and Russian languages, both texts being equally authentic.

For the United States of America:
Richard Nixon
President of the United States of America

For the Union of Soviet Socialist Republics:
L. I. Brezhnev
General Secretary of the Central Committee of the CPSU

65. SALT II Treaty between the United States of America and the Union of Soviet Socialist Republics on the Limitation of Strategic Offensive Arms, Vienna, June 18, 1979 (including Protocol and Agreed Understandings)

The United States of America and the Union of Soviet Socialist Republics, hereinafter referred to as the Parties,

Conscious that nuclear war would have devastating consequences for all mankind,

Proceeding from the Basic Principles of Relations Between the United States of America and the Union of Soviet Socialist Republics of May 29, 1972,

Attaching particular significance to the limitation of strategic arms and determined to continue their efforts begun with the Treaty on the Limitation of Anti-Ballistic Missile Systems and the Interim Agreement on Certain Measures with Respect to the Limitation of Strategic Offensive Arms, of May 26, 1972,

Convinced that the additional measures limiting strategic offensive arms provided for in this Treaty will contribute to the improvement of relations between the Parties, help to reduce the risk of outbreak of nuclear war, and strengthen international peace and security,

Mindful of their obligations under Article VI of the Treaty on the Non-Proliferation of Nuclear Weapons,

Guided by the principle of equality and equal security,

Recognizing that the strengthening of strategic stability meets the interests of the Parties and the interests of international security,

Reaffirming their desire to take measures for the further limitation and for the further reduction of strategic arms, having in mind the goal of achieving general and complete disarmament,

Declaring their intention to undertake in the near future negotiations further to limit and further to reduce strategic offensive arms,

Have agreed as follows:

Article I

Each Party undertakes, in accordance with the provisions of this Treaty, to limit strategic offensive arms quantitatively and qualitatively, to exercise

This document appeared in United States Arms Control and Disarmament Agency, *Arms Control and Disarmament Agreements* (Washington, D.C.: Government Printing Office 1982), pp. 246–69.

restraint in the development of new types of strategic offensive arms, and to adopt other measures provided for in this Treaty.

Article II

For the purposes of this Treaty:

1. Intercontinental ballistic missile (ICBM) launchers are land-based launchers of ballistic missiles capable of a range in excess of the shortest distance between the northeastern border of the continental part of the territory of the United States of America and the northwestern border of the continental part of the territory of the Union of Soviet Socialist Republics, that is, a range in excess of 5,500 kilometers.

2. Submarine-launched ballistic missile (SLBM) launchers are launchers of ballistic missiles installed on any nuclear-powered submarine or launchers of modern ballistic missiles installed on any submarine, regardless of its type.

3. Heavy bombers are considered to be:

(a)   currently, for the United States of America, bombers of the B-52 and B-1 types, and for the Union of Soviet Socialist Republics, bombers of the Tupolev-95 and Myasishchev types;
(b)   in the future, types of bombers which can carry out the mission of a heavy bomber in a manner similar or superior to that of bombers listed in subparagraph (a) above;
(c)   types of bombers equipped for cruise missiles capable of a range in excess of 600 kilometers; and
(d)   types of bombers equipped for ASBMs.

4. Air-to-surface ballistic missiles (ASBMs) are any such missiles capable of a range in excess of 600 kilometers and installed in an aircraft or on its external mountings.

5. Launchers of ICBMs and SLBMs equipped with multiple independently targetable reentry vehicles (MIRVs) are launchers of the types developed and tested for launching ICBMs or SLBMs equipped with MIRVs.

6. ASBMs equipped with MIRVs are ASBMs of the types which have been flight-tested with MIRVs.

7. Heavy ICBMs are ICBMs which have a launch-weight greater or a throw-weight greater than that of the heaviest, in terms of either launch-weight or throw-weight, respectively, of the light ICBMs deployed by either Party as of the date of signature of this Treaty.

8. Cruise missiles are unmanned, self-propelled, guided, weapon-

delivery vehicles which sustain flight through the use of aerodynamic lift over most of their flight path and which are flight-tested from or deployed on aircraft, that is, air-launched cruise missiles, or such vehicles which are referred to as cruise missiles in subparagraph I(b) of Article IX.

Article III

1. Upon entry into force of this Treaty, each Party undertakes to limit ICBM launchers, SLBM launchers, heavy bombers, and ASBMs to an aggregate number not to exceed 2,400.

2. Each Party undertakes to limit, from January 1, 1982, strategic offensive arms referred to in paragraph 1 of this Article to an aggregate number not to exceed 2,250 and to initiate reductions of those arms which as of that date would be in excess of this aggregate number.

3. Within the aggregate numbers provided for in paragraphs 1 and 2 of this Article and subject to the provisions of this Treaty, each Party has the right to determine the composition of these aggregates.

4. For each bomber of a type equipped for ASBMs, the aggregate number provided for in paragraphs 1 and 2 of this Article shall include the maximum number of such missiles for which a bomber of that type is equipped for one operational mission.

5. A heavy bomber equipped only for ASBMs shall not itself be included in the aggregate numbers provided for in paragraphs 1 and 2 of this Article.

6. Reductions of the numbers of strategic offensive arms required to comply with the provisions of paragraphs 1 and 2 of this Article shall be carried out as provided for in Article XI.

Article IV

1. Each Party undertakes not to start construction of additional fixed ICBM launchers.

2. Each Party undertakes not to relocate fixed ICBM launchers.

3. Each Party undertakes not to convert launchers of light ICBMs, or of ICBMs of older types deployed prior to 1964, into launchers of heavy ICBMs of types deployed after that time.

4. Each party undertakes in the process of modernization and replacement of ICBM silo launchers not to increase the original internal volume of an ICBM silo launcher by more than thirty-two percent. Within this limit each Party has the right to determine whether such an increase will be made through an increase in the original diameter or in the original depth of an ICBM silo launcher, or in both of these dimensions.

5. Each Party undertakes:

(a)   not to supply ICBM launcher deployment areas with intercontinental ballistic missiles in excess of a number consistent with normal deployment, maintenance, training, and replacement requirements;

(b)   not to provide storage facilities for or to store ICBMs in excess of normal deployment requirements at launch sites of ICBM launchers;

(c)   not to develop, test, or deploy systems for rapid reload of ICBM launchers.

6. Subject to the provisions of this Treaty, each Party undertakes not to have under construction at any time strategic offensive arms referred to in paragraph 1 of Article III in excess of numbers consistent with a normal construction schedule.

7. Each Party undertakes not to develop, test, or deploy ICBMs which have a launch-weight greater or a throw-weight greater than that of the heaviest, in terms of either launch-weight or throw-weight, respectively, of the heavy ICBMs deployed by either Party as of the date of signature of this Treaty.

8. Each Party undertakes not to convert land-based launchers of ballistic missiles which are not ICBMs into launchers for launching ICBMs, and not to test them for this purpose.

9. Each Party undertakes not to flight-test or deploy new types of ICBMs, that is, types of ICBMs not flight-tested as of May 1, 1979, except that each Party may flight-test and deploy one new type of light ICBM.

10. Each Party undertakes not to flight-test or deploy ICBMs of a type flight-tested as of May 1, 1979 with a number of reentry vehicles greater than the maximum number of reentry vehicles with which an ICBM of that type has been flight-tested as of that date.

11. Each Party undertakes not to flight-test or deploy ICBMs of the one new type permitted pursuant to paragraph 9 of this Article with a number of reentry vehicles greater than the maximum number of reentry vehicles with which an ICBM of either Party has been flight-tested as of May 1, 1979, that is ten.

12. Each Party undertakes not to flight-test or deploy SLBMs with a number of reentry vehicles greater than the maximum number of reentry vehicles with which an SLBM of either Party has been flight-tested as of May 1, 1979, that is, fourteen.

13. Each Party undertakes not to flight-test or deploy ASBMs with a number of reentry vehicles greater than the maximum number of reentry

vehicles with which an ICBM of either Party has been flight-tested as of May 1, 1979, that is, ten.

14. Each Party undertakes not to deploy at any one time on heavy bombers equipped for cruise missiles capable of a range in excess of 600 kilometers a number of such cruise missiles which exceeds the product of 28 and the number of such heavy bombers. . . .

Soviet Backfire Statement

On June 16, 1979, President Brezhnev handed President Carter the following written statement:

"The Soviet side informs the U.S. side that the Soviet 'Tu-22M' [alternatively designated Tu-29] airplane, called 'Backfire' in the U.S.A., is a medium-range bomber and that it does not intend to give this airplane the capability of operating at intercontinental distances. In this connection, the Soviet side states that it will not increase the radius of action of this airplane in such a way as to enable it to strike targets on the territory of the U.S.A. Nor does it intend to give it such a capability in any other manner, including by in-flight refueling. At the same time, the Soviet side states that it will not increase the production rate of this airplane as compared to the present rate."

President Brezhnev confirmed that the Soviet Backfire production rate would not exceed 30 per year.

President Carter stated that the United States enters into the SALT II agreement on the basis of the commitments contained in the Soviet statement and that it considers the carrying out of these commitments to be essential to the obligation assumed under the treaty.

Memorandum of Understanding between the United States of America and the Union of Soviet Socialist Republics Regarding the Establishment of a Data Base on the Numbers of Strategic Offensive Arms

For the purposes of the Treaty between the United States of America and the Union of Soviet Socialist Republics on the Limitation of Strategic Offensive Arms, the Parties have considered data on numbers of strategic offensive arms and agree that as of November 1, 1978 there existed the following numbers of strategic offensive arms subject to the limitations provided for in the Treaty which is being signed today.

	U.S.A.	U.S.S.R.
Launchers of ICBMs	1,054	1,398
Fixed launchers of ICBMs	1,054	1,398
Launchers of ICBMs equipped with MIRVs	550	576
Launchers of SLBMs	656	950
Launchers of SLBMs equipped with MIRVs	496	128
Heavy bombers	574	156
Heavy bombers equipped for cruise missiles capable of a range in excess of 600 kilometers	0	0
Heavy bombers equipped only for ASBMs	0	0
ASBMs	0	0
ASBMs equipped with MIRVs	0	0

At the time of entry into force of the Treaty the Parties will update the above agreed data in the categories listed in this Memorandum.

Done at Vienna on June 18, 1979, in two copies, each in the English and Russian languages, both texts being equally authentic.

FOR THE UNITED STATES OF AMERICA

FOR THE UNION OF SOVIET SOCIALIST REPUBLICS

CHIEF OF THE UNITED STATES DELEGATION TO THE STRATEGIC ARMS LIMITATION TALKS

CHIEF OF THE U.S.S.R. DELEGATION TO THE STRATEGIC ARMS LIMITATION TALKS

Statement of Data on the Numbers of Strategic Offensive Arms of the Date of Signature of the Treaty

The United States of America declares that as of June 18, 1979 it possesses the following numbers of strategic offensive arms subject to the limitations provided for in the Treaty which is being signed today.

Launchers of ICBMs	1,054
Fixed launchers of ICBMs	1,054
Launchers of ICBMs equipped with MIRVs	550
Launchers of SLBMs	656
Launchers of SLBMs equipped with MIRVs	496
Heavy bombers	573
Heavy bombers equipped for cruise missiles capable of a range in excess of 600 kilometers	3
Heavy bombers equipped only for ASBMs	0

ASBMs   0
ASBMs equipped with MIRVs   0

June 18, 1979

                                        CHIEF OF THE
                                        UNITED STATES DELEGATION
                                        TO THE STRATEGIC ARMS
                                        LIMITATION TALKS

Statement of Data on the Numbers of Strategic Offensive Arms of the Date of Signature of the Treaty

The Union of Soviet Socialist Republics declares that as of June 18, 1979 it possesses the following numbers of strategic offensive arms subject to the limitations provided for in the Treaty which is being signed today:

Launchers of ICBMs	1,398
Fixed launchers of ICBMs	1,398
Launchers of ICBMs equipped with MIRVs	608
Launchers of SLBMs	950
Launchers of SLBMs equipped with MIRVs	144
Heavy bombers	156
Heavy bombers equipped for cruise missiles capable of a range in excess of 600 kilometers	0
Heavy bombers equipped only for ASBMs	0
ASBMs	0
ASBMs equipped with MIRVs	0

June 18, 1979

                                        CHIEF OF THE
                                        U.S.S.R. DELEGATION
                                        TO THE STRATEGIC ARMS
                                        LIMITATION TALKS

Joint Statement of Principles and Basic Guidelines for Subsequent Negotiations on the Limitation of Strategic Arms

The United States of America and the Union of Soviet Socialist Republics, hereinafter referred to as the Parties,

Having concluded the Treaty on the Limitation of Strategic Offensive Arms,

Reaffirming that the strengthening of strategic stability meets the interests of the Parties and the interests of international security,

Convinced that early agreement on the further limitation and further reduction of strategic arms would serve to strengthen international peace and security and to reduce the risk of outbreak of nuclear war,

Have agreed as follows:

First. The Parties will continue to pursue negotiations, in accordance with the principle of equality and equal security, on measures for the further limitation and reduction in the numbers of strategic arms, as well as for their further qualitative limitation.

In furtherance of existing agreements between the Parties on the limitation and reduction of strategic arms, the Parties will continue, for the purposes of reducing and averting the risk of outbreak of nuclear war, to seek measures to strengthen strategic stability by, among other things, limitations on strategic offensive arms most destabilizing to the strategic balance and by measures to reduce and to avert the risk of surprise attack.

Second. Further limitations and reductions of strategic arms must be subject to adequate verification by national technical means, using additionally, as appropriate, cooperative measures contributing to the effectiveness of verification by national technical means. The Parties will seek to strengthen verification and to perfect the operation of the Standing Consultative Commission in order to promote assurance of compliance with the obligations assumed by the parties.

Third. The Parties shall pursue in the course of these negotiations, taking into consideration factors that determine the strategic situation, the following objectives:

1) significant and substantial reductions in the numbers of strategic offensive arms;

2) qualitative limitations on strategic offensive arms, including restrictions on the development, testing, and deployment of new types of strategic offensive arms and on the modernization of existing strategic offensive arms;

3) resolution of the issues included in the Protocol to the Treaty Between the United States of America and the Union of Soviet Socialist Republics on the Limitation of Strategic Offensive Arms in the context of the negotiations relating to the implementation of the principles and objectives set out herein.

Fourth. The Parties will consider other steps to ensure and enhance strategic stability, to ensure the equality and equal security of the Parties, and to implement the above principles and objectives. Each Party will be free to raise any issue relative to the further limitation of strategic arms. The Parties will also consider further joint measures, as appropriate, to strengthen international peace and security and to reduce the risk of outbreak of nuclear war.

Vienna, June 18, 1979

FOR THE UNITED STATES OF AMERICA	FOR THE UNION OF SOVIET SOCIALIST REPUBLICS
PRESIDENT OF THE UNITED STATES OF AMERICA	GENERAL SECRETARY OF THE CPSU, CHAIRMAN OF THE PRESIDIUM OF THE SUPREME SOVIET OF THE USSR

66. U.S. Bishops' Pastoral Letter on War and Peace: A Summary

The Second Vatican Council opened its evaluation on modern warfare with the statement: "The whole human race faces a moment of supreme crisis in its advance toward maturity." We agree with the council's assessment; the crisis of the moment is embodied in the threat which nuclear weapons pose for the world and much that we hold dear in the world. We have seen and felt the effects of the crisis of the nuclear age in the lives of people we serve. Nuclear weaponry has drastically changed the nature of warfare and the arms race poses a threat to human life and human civilization which is without precedent.

We write this letter from the perspective of Catholic faith. Faith does not insulate us from the daily challenges of life, but intensifies our desire to address them precisely in light of the gospel which has come to us in the person of the Risen Christ. Through the resources of faith and reason we desire in this letter to provide hope for people in our day and direction toward a world freed of the nuclear threat.

As Catholic bishops we write this letter as an exercise of our teaching ministry. The Catholic tradition on war and peace is a long and complex one; it stretches from the Sermon on the Mount to the statements of Pope John Paul II. We wish to explore and explain the resources of the moral-religious teaching and to apply it to specific questions of our day. In doing this we realize and we want readers of this letter to recognize that not all statements in this letter have the same moral authority. At times we state universally binding moral principles found in the teaching of the church; at other times the pastoral letter makes specific applications, observations and recommendations which allow for diversity of opinion on the part of those who assess the factual data of situations differently. However, we expect Catholics to give our moral judgments serious consideration when they are forming their own views on specific problems.

The experience of preparing this letter has manifested to us the range of strongly held opinion in the Catholic community on questions of both fact and judgment concerning issues of war and peace. We urge mutual respect among individuals and groups in the church as this letter is analyzed and discussed. Obviously, as bishops, we believe that such differences should be expressed within the framework of Catholic moral teaching. We need in the church not only conviction and commitment, but also civility and charity.

While this letter is addressed principally to the Catholic community, we want it to make a contribution to the wider public debate in our country on the dangers and dilemmas of the nuclear age. Our contribution will not be primarily technical or political, but we are convinced that there is no satisfactory answer to the human problems of the nuclear age which fails to consider the moral and religious dimensions of the questions we face.

Although we speak in our own name as Catholic bishops of the church in the United States, we have been conscious in the preparation of this letter of the consequences our teaching will have not only for the United States, but for other nations as well. One important expression of this awareness has been the consultation we have had, by correspondence and in an important meeting held at the Vatican (Jan. 18–19, 1983), with representatives of European bishops' conferences. This consultation with bishops of other countries and, of course, with the Holy See has been very helpful to us.

Catholic teaching has always understood peace in positive terms. In the words of Pope John Paul II: "Peace is not just the absence of war. . . . Like a cathedral, peace must be constructed patiently and with unshakable faith" (Coventry, England, 1982). Peace is the fruit of order. Order in human society must be shaped on the basis of respect for the transcendence of God and the unique dignity of each person, understood in terms of freedom, justice, truth and love. To avoid war in our day we must be intent on building peace in an increasingly interdependent world. In Part III of this letter we set forth a positive vision of peace and the demands such a vision makes on diplomacy, national policy, and personal choices.

While pursuing peace incessantly, it is also necessary to limit the use of force in a world comprised of nation states, faced with common problems but devoid of an adequate international political authority. Keeping the peace in the nuclear age is a moral and political imperative. In Parts I and II of this letter we set forth both the principles of Catholic teaching on war and a series of judgments, based on these principles, about concrete policies. In making these judgments we speak as moral teachers, not as technical experts.

I. Some Principals, Norms and Premises of Catholic Teaching:

A. *On War:*

1. Catholic teaching begins in every case with a presumption against war and for peaceful settlement of disputes. In exceptional cases, determined by the moral principles of the Just War tradition, some uses of force are permitted.

2. Every nation has a right and duty to defend itself against unjust aggression.

3. Offensive war of any kind is not morally justifiable.

4. It is never permitted to direct nuclear or conventional weapons to "the indiscriminate destruction of whole cities or vast areas with their populations . . ." ("The Pastoral Constitution on the Church in the Modern World," No. 80). The intentional killing of innocent civilians or noncombatants is always wrong.

5. Even defensive response to unjust attack can cause destruction which violates the principle of proportionality, going far beyond the limits of legitimate defense. This judgment is particularly important when assessing planned use of nuclear weapons. No defensive strategy, nuclear or conventional which exceeds the limits of proportionality is morally permissible.

B. *On Deterrence:*

1. "In current conditions 'deterrence' based on balance, certainly not as an end in itself but as a step on the way toward a progressive disarmament,

may still be judged morally acceptable. Nonetheless, in order to ensure peace, it is indispensable not to be satisfied with this minimum which is always susceptible to the real danger of explosion." (Pope John Paul II, Message to U.N. Special Session on Disarmament, No. 8; June 1982.)

2. No use of nuclear weapons which would violate the principles of discrimination or proportionality may be intended in a strategy of deterrence. The moral demands of Catholic teaching require resolute willingness not to intend or to do moral evil even to save our own lives or the lives of those we love.

3. Deterrence is not an adequate strategy as a long-term basis for peace; it is a transitional strategy justifiable only in conjunction with resolute determination to pursue arms control and disarmament. We are convinced that "the fundamental principle on which our present peace depends must be replaced by another, which declares that the true and solid peace of nations consists not in equality of arms but in mutual trust alone." (Pope John XXIII, "Peace On Earth," No. 113.)

C. The Arms Race and Disarmament:

1. The arms race is one of the greatest curses on the human race; it is to be condemned as a danger, an act of aggression against the poor, and a folly which does not provide the security it promises. (Cf: "The Pastoral Constitution . . . ," No. 81; Statement of the Holy See to the United Nations, 1976.)

2. Negotiations must be pursued in every reasonable form possible; they should be governed by the "demand that the arms race should cease; that the stockpiles which exist in various countries should be reduced equally and simultaneously by the parties concerned; that nuclear weapons should be banned; and that a general agreement should eventually be reached about progressive disarmament and an effective method of control." (Pope John XXIII, "Peace On Earth," No. 112.)

D. On Personal Conscience:

1. Military Service: "All those who enter the military service in loyalty to their country should look upon themselves as the custodians of the security and freedom of their fellow countrymen; and when they carry out their duty properly, they are contributing to the maintenance of peace." ("The Pastoral Constitution . . . ," No. 79.)

2. Conscientious Objection: "Moreover, it seems just that laws should make humane provision for the case of conscientious objectors who refuse to carry arms, provided they accept some other form of community service." ("The Pastoral Constitution . . . ," No. 79.)

3. Non-violence: "In this same spirit we cannot but express our admiration for all who forgo the use of violence to vindicate their rights and resort to other means of defense which are available to weaker parties, provided it can be done without harm to the rights and duties of others and of the community." ("The Pastoral Constitution . . . ," No. 78.)

4. Citizens and Conscience: "Once again we deem it opportune to remind our children of their duty to take an active part in public life, and to contribute towards the attainment of the common good of the entire human family as well as to that of their own political community. . . . In other words, it is necessary that human beings, in the intimacy of their own consciences, should so live and act in their temporal lives as to create a synthesis between scientific, technical and professional elements on the one hand, and spiritual values on the other." (Pope John XXIII, "Peace On Earth," No. 146;150.)

II. Moral Principles and Policy Choices:

As bishops in the United States, assessing the concrete circumstances of our society, we have made a number of observations and recommendations in the process of applying moral principles to specific policy choices.

A. *On the Use of Nuclear Weapons:*
1. Counter-Population Use: Under no circumstances may nuclear weapons or other instruments of mass slaughter be used for the purpose of destroying population centers or other predominantly civilian targets. Retaliatory action which would indiscriminately and disproportionately take many wholly innocent lives, lives of people who are in no way responsible for reckless actions of their government, must also be condemned.

2. The Initiation of Nuclear War: We do not perceive any situation in which the deliberate initiation of nuclear war, on however restricted a scale, can be morally justified. Non-nuclear attacks by another state must be resisted by other than nuclear means. Therefore, a serious obligation exists to develop morally acceptable non-nuclear defensive strategies as rapidly as possible. In this letter we urge NATO to move rapidly toward the adoption of a "no first use" policy but we recognize this will take time to implement and will require the development of an adequate alternative defense posture.

3. Limited Nuclear War: Our examination of the various arguments on this question makes us highly skeptical about the real meaning of "limited." One of the criteria of the just war teaching is that there must be a

reasonable hope of success in bringing about justice and peace. We must ask whether such a reasonable hope can exist once nuclear weapons have been exchanged. The burden of proof remains on those who assert that meaningful limitation is possible. In our view the first imperative is to prevent any use of nuclear weapons and we hope that leaders will resist the notion that nuclear conflict can be limited, contained or won in any traditional sense.

B. On Deterrence:

In concert with the evaluation provided by Pope John Paul II, we have arrived at a strictly conditioned moral acceptance of deterrence. In this letter we have outlined criteria and recommendations which indicate the meaning of conditional acceptance of deterrence policy. We cannot consider such a policy adequate as a long-term basis for peace.

C. On Promoting Peace:

1. We support immediate, bilateral, verifiable agreements to halt the testing, production and deployment of new nuclear weapons systems. This recommendation is not to be identified with any specific political initiative.

2. We support efforts to achieve deep cuts in the arsenals of both superpowers; efforts should concentrate first on systems which threaten the retaliatory forces of either major power.

3. We support early and successful conclusion of negotiations of a comprehensive test ban treaty.

4. We urge new efforts to prevent the spread of nuclear weapons in the world, and to control the conventional arms race, particularly the conventional arms trade.

5. We support, in an increasingly interdependent world, political and economic policies designed to protect human dignity and to promote the human rights of every person, especially the least among us. In this regard, we call for the establishment of some form of global authority adequate to the needs of the international common good.

This letter includes many judgments from the perspective of ethics, politics, and strategy needed to speak concretely and correctly to the "moment of supreme crisis" identified by Vatican II. We stress again that readers should be aware, as we have been, of the distinction between our statement of moral principles and of official Church teaching and our application of these to concrete issues. We urge that special care be taken not to use passages out of context; neither should brief portions of this document be cited to support positions it does not intend to convey or which are not truly in accord with the spirit of its teaching.

In concluding this summary we respond to two key questions often asked about this pastoral letter:

Why do we address these matters fraught with such complexity, controversy and passion? We speak as pastors, not politicians. We are teachers, not technicians. We cannot avoid our responsibility to lift up the moral dimensions of the choices before our world and nation. The nuclear age is an era of moral as well as physical danger. We are the first generation since Genesis with the power to threaten the created order. We cannot remain silent in the face of such danger. Why do we address these issues? We are simply trying to live up to the call of Jesus to be peacemakers in our own time and situation.

What are we saying? Fundamentally, we are saying that the decisions about nuclear weapons are among the most pressing moral questions of our age. While these decisions have obvious military and political aspects, they involve fundamental moral choices. In simple terms, we are saying that good ends (defending one's country, protecting freedom, etc.) cannot justify immoral means (the use of weapons which kill indiscriminately and threaten whole societies). We fear that our world and nation are headed in the wrong direction. More weapons with greater destructive potential are produced every day. More and more nations are seeking to become nuclear powers. In our quest for more and more security we fear we are actually becoming less and less secure.

In the words of our Holy Father, we need a "moral about-face." The whole world must summon the moral courage and technical means to say no to nuclear conflict; no to weapons of mass destruction; no to an arms race which robs the poor and the vulnerable; and no to the moral danger of a nuclear age which places before humankind indefensible choices of constant terror or surrender. Peacemaking is not an optional commitment. It is a requirement of our faith. We are called to be peacemakers, not by some movement of the moment but by our Lord Jesus. The content and context of our peacemaking is set not by some political agenda or ideological program, but by the teaching of his Church.

Ultimately, this letter is intended as an expression of Christian faith, affirming the confidence we have that the Risen Lord remains with us precisely in moments of crisis. It is our belief in his presence and power among us which sustains us in confronting the awesome challenge of the nuclear age. We speak from faith to provide hope for all who recognize the challenge and are working to confront it with the resources of faith and reason.

To approach the nuclear issue in faith is to recognize our absolute need for prayer: We urge and invite all to unceasing prayer and works of penance for peace with justice for all people. In a spirit of prayerful hope we present this message of peace.

---

67. William P. Clark, National Security Council, to Archbishop Joseph L. Bernadin, Chairman of the National Conference of Catholic Bishops' Committee on War and Peace, November 16, 1982

I would like to take this opportunity to respond, on behalf of President Reagan, Secretary Shultz, Secretary Weinberger, Director Rostow, and other Administration officials, to the request for our views on the second draft of the pastoral letter recently prepared by the Ad Hoc Committee on War and Peace for review by the National Conference of Catholic Bishops. Let me assure you that we have read these drafts with great interest and care.

I believe we can agree that the purpose of any moral theory of defense is "not, in the first place, to legitimize war, but to prevent it," and this, of course, is what American deterrence policy is designed to achieve. I believe we can also agree that any proposed change in strategic systems or doctrines, as well as any recommendation, whether proposed by the U.S. Government or by your committee, should be judged "in light of whether it will render steps toward arms control and disarmament more, or less, likely." We believe that our weapons systems (which are not designed to be "first-strike" systems), our deterrence posture (which is defensive), and our arms control initiatives (which call for deep and verifiable reductions) do conform to these objectives.

As with the committee's first draft, I am especially troubled in reading the second draft of the pastoral letter to find none of the serious U.S. arms control efforts, including major initiatives and ongoing U.S.-Soviet negotiations, described or even noted in the text. Ours are not proposals for freezes on current high ceilings. Such freezes would remove incentives for achieving reductions and would, in any case, require extensive prior negotiations to reach agreement on what numbers and systems to freeze, and on how such freezes might be effectively verified. Ours are initiatives for

This essay appeared in the *New York Times*, November 17, 1982, p. 11.

reduction, or even elimination, of the most destabilizing systems. They involve new verification and confidence-building measures designed both to build trust and to assure compliance.

Because these important initiatives and negotiations have again been ignored in the draft pastoral letter, although they so clearly conform to the hopes of all concerned with reducing the arsenals and the risks of war and promoting the path of peace, I would like to summarize them for you again. I do so with a renewed hope that the comments your committee receives from U.S. Government officials in response to the committee's requests will be carefully considered, just as your committee asks that its draft letter "receive a respectful consideration" from others.

This Administration's arms control efforts include the following major initiatives:

In the U.S.-Soviet negotiations on strategic arms (START), which began on June 30, 1982, we are proposing to begin with a one-third reduction in the number of warheads on the land and sea-based ballistic missiles and a reduction in the most destabilizing systems of all, the land-based ballistic missiles, to about one-half of the current U.S. levels. In a second phase, we propose to reduce the destructive potential of the remaining missiles to equal levels, lower than we now have, and we could include other strategic systems as well.

In the U.S.-Soviet negotiations on intermediate-range nuclear forces (INF), which began on Nov. 30, 1981, we have proposed to begin with the total elimination of the forces considered the most destabilizing and threatening by both sides, the land-based missile systems. We and our NATO allies have offered to cancel plans for the deployment of U.S. Pershing and ground-launched cruise missiles in exchange for the corresponding destruction of Soviet SS-20, SS-4 and SS-5 missiles. Other elements of the balance could be limited subsequently.

In the multilateral negotiations on mutual and balanced force reductions (MBFR), the U.S. and its NATO allies are proposing to the Warsaw Pact nations major initial reductions in military personnel to common ceilings and a wide range of new verification measures.

In the areas of limiting nuclear testing and chemical and biological weapons, the U.S. is actively participating in discussions in the Committee on Disarmament in Geneva to develop the verification and compliance procedures that would make such limitations truly effective. We are, of course, particularly distressed by the extensive and inhuman use by the Soviet Union and its allies of toxins and chemicals against the defenseless populations of Afghanistan, Laos and Cambodia.

In all of our ongoing arms control negotiations and discussions, we are emphasizing the importance of substantial early reductions and of effective verification and confidence-building measures. Your committee will surely recognize that the Administration's nuclear reductions proposals clearly conform to the pastoral letter's recommendations for cuts in nuclear arsenals, and that the other multilateral efforts in which we are currently engaged conform closely with the letter's call for efforts "aimed at reducing and limiting conventional forces and at building confidence between possible adversaries, especially in regions of major military confrontation, as well as those addressed to outlawing effectively the use of chemical and biological weapons."

I continue to believe that as the Bishops Conference reviews new drafts of the pastoral letter, a clear presentation of these initiatives should lead to the Bishops Conference's strong support for them. As I noted in my comments on the first draft, such support would prove enormously helpful in making clear to the world America's seriousness in our efforts and would, in particular, add to Soviet incentives to agree to the reductions and verifiable limitations that we are seeking.

IX

Nuclear Power

*E*vents between 1951 and 1954 accelerated the drive for commercial nuclear power. In the winter of 1951, a small government reactor at the Idaho National Engineering Laboratory produced electricity for the first time. Eisenhower's "Atoms for Peace" speech in December 1953 emphasized an international program of civilian uses of atomic energy, especially power reactors for export to fuel-hungry Europe.

That same December, the Atomic Energy Commission selected the Duquesne Light and Power Company, a Pittsburgh utility, to join a three-way partnership in the construction and operation of the first civilian nuclear power plant. Duquesne would provide the land and about $300 million and would operate and maintain the plant. Duquesne's partners would be the federal government and the Westinghouse corporation, which would manufacture the reactor. "Off the shelf" technology would be used. The reactor was a civilian version of a pressurized water reactor Westinghouse was building for Hyman G. Rickover's Navy program. Admiral Lewis L. Strauss, a science adviser to Eisenhower who would later chair the Atomic Energy Commission, recommended adapting the Navy's pressurized water reactor to civilian purposes, otherwise "much time and momentum would be lost by turning over this project to a new contractor. . . ." To act quickly was paramount.

Congress encouraged further private reactor development when it amended the Atomic Energy Act of 1946 in the summer of 1954. Under the 1954 Act, free enterprise would provide the impetus for civilian nuclear power. The new legislation permitted the government to license private corporations to build and operate nuclear-fueled power plants. The Atomic Energy Commission could continue to fund research and development, but *the act prohibited the Commission from constructing power "reactors for the purpose of selling or distributing electricity." Industry, so Congress and the Eisenhower administration believed, could do it better. "The goal of atomic power at competitive prices will be reached more quickly if private enterprise, using private funds, is . . . encouraged to play a far larger role in the development of atomic power. . . ." That role, as the congressmen perceived it, would be plant construction, while the Atomic Energy Commission would continue to subsidize research and development projects shunned by industry.*

Nuclear power advocates exuded optimism in 1954. The U.S.S. Nautilus, the world's first nuclear-powered submarine, was launched that year. The Atomic Energy Act had been amended to provide more government assistance for private development of nuclear energy. President Eisenhower continued to stress the peaceful uses of the atom. The announcement of the construction of a nuclear power plant at Shippingport, Pennsylvania, by the Atomic Energy Commission and Duquesne Electric Company and a five-year commercial reactor demonstration program highlighted this emphasis. Ike predicted the day when the atom "will be wholly devoted to the constructive purposes of man." Congressmen spoke about "cheap power" within a decade. Utilities, experts argued, would save billions of dollars. The president of General Electric estimated that 50 percent of the new electric power plants would be atomic by 1976. And Lewis Strauss predicted that "our children will enjoy in their homes electrical energy too cheap to meter."

By the winter of 1954–55, the Atomic Energy Commission and several utility company consortiums were ready to demonstrate the feasibility of nuclear power. In December the Atomic Energy Commission announced the Power Demonstration Reactor Program and asked for proposals from the power companies by April 1, 1955. The Atomic Energy Commission hoped to encourage industry activity with government support limited to research and development. At the same time, the commission sought to extend technical knowledge about a wide range of reactor types and, in the end, make nuclear power economically feasible. Four groups replied. Detroit Edison proposed a fast-breeder reactor; Yankee Atomic, a consortium of thirteen New England utilities, suggested a pressurized water reactor on the Shippingport mode; the Nuclear Power Group, headed by Commonwealth Edison in Chicago, proposed a boiling water reactor; and the Consumers Public Power District in Nebraska submitted plans for a sodium-graphite reactor.

In addition, one additional utility offered to build its own nuclear plant without any outside assistance either from the government or in consort with other companies. The Consolidated Edison Company of New York, managed by men steeped in the tradition of private enterprise, submitted an application to construct a nuclear plant at Indian Point, some thirty-five miles up the Hudson River from New York City. By May 1956 the Atomic Energy Commission had granted Con Edison's plans for a pressurized water reactor, designed by Babcock & Wilcox, the first public power construction permit.

In encouraging private development of nuclear power, the Atomic Energy Commission had created a new area of regulatory concerns. Government reactors had always been located in remote areas. But to give nuclear energy a chance to be economically feasible, power companies sought to locate plants near their customers and existing power grids. Therefore, the commission had to draw up new siting plans, and radiation standards had to ensure the public's health and safety. A dilemma emerged.

The Atomic Energy Act of 1954 gave the Atomic Energy Commission responsibility for regulating and licensing power reactors and, at the same time, the task of promoting nuclear power. Both the regulatory and non-regulatory functions were held by the Division of Civilian Application (the Division of Military Application was responsible for weapons) until 1957 when a separate regulatory staff was established. In addition, an Advisory Committee on Reactor Safeguards, a group of fifteen experts, provided advice to the Commission on the safety and licensing of nuclear reactors. The Advisory Committee, nicknamed the "Brake Department," spoke out at times against the commission's policy on safety, especially in its concern about Detroit Edison's Fermi breeder reactor. In 1974, Congress emphasized this functional split when it abolished the Atomic Energy Commission and established the Nuclear Regulatory Commission to regulate nuclear power and the Energy Research and Development Administration (later the Department of Energy) to promote its use.

Shippingport began producing electricity for commercial use in 1957. Commonwealth Edison's Dresden plant near Chicago began operating in 1960, the Yankee Atomic plant near Rowe, Massachusetts, went on line in 1961, and Con Edison's Indian Point plant started up in 1962. By late 1970, the nuclear power industry had eighteen operating nuclear plants and had applications pending for twenty more. By 1975, fifty-five plants were in operation and one hundred seventy-eight more were planned or under construction.

By the mid-1960s, experimentation over power reactor types had ceased. The pressurized water reactor, built by Westinghouse and other corporations, and the boiling water reactor built by General Electric became the industry standards, and plant orders increased. Industry spokesmen be-

lieved over 50 percent of the nation's electric power would be nuclear by the year 2000.

The glowing promise of nuclear energy, however, was never fulfilled. Opposition to a proposed nuclear plant in New York City in early 1963 marked the beginnings of public concern. Widespread focus on the danger of fallout, the moratorium on atmospheric testing of nuclear weapons, and President John F. Kennedy's initiatives toward a test-ban treaty contributed to linking nuclear weapons with nuclear power plants in the public perception.

In addition, the Union of Concerned Scientists and other groups feared the increased size of the nuclear reactors and the ability of the industry to make them safe enough to contain a serious accident. Consequently, opponents to nuclear power led a growing number of attempts to block the construction or licensing of new atomic power stations. At public hearings, these groups intervened to delay or halt plant operation for reasons other than radiation safety. A proposed plant in Easton, New York, for example, failed to gain approval because it was located too close to the Saratoga National Historical Park and received adverse comments from the Advisory Council on Historic Preservation. A planned reaction in Oregon failed to win voter approval in a special referendum that restricted the use of its public funding.

The National Environmental Policy Act of 1969 became the chief weapon in changing the Atomic Energy Commission's licensing procedures. In the Calvert Cliffs decision in 1971, the federal courts ruled that the commission had to consider the total impact a nuclear plant would have on the environment, including water quality, and perform a cost-benefit analysis on the economic, technical, and other benefits against alternatives, including not building the plant. The decision restricted the commission's licensing program and slowed down the planning and construction of nuclear power plants.

Even as the cost of oil soared after the 1973 boycott, the cost of nuclear plants also rose. Many in the planning and construction stage could no longer be competitive as building costs doubled, then tripled, as legal delays and inflation pushed costs higher. By 1976, utility orders for new nuclear plants had already begun to decline, several years before the accident at Three Mile Island seemed to doom nuclear power in the public's perception.

The debate over nuclear power intensified. The 1979 Three Mile Island accident provided new ammunition for opponents who sought to use the accident to close down all atomic plants. Industry proponents argued that the containment system at Three Mile Island worked, in spite of human errors that compounded the problem. Moreover, they said, the steady depletion of fossil fuels would make nuclear power inevitable. The peaceful atom continues to stir as much controversy as the bomb.

The best short history of the development of the nuclear power industry is by Irwin Bupp and J. C. Derian, Light Water: How the Nuclear Dream Dissolved *(New York: Basic Books, 1978). A more technical study is Frank G. Dawson,* Nuclear Power: Development and Management of a Technology *(Seattle and London: University of Washington Press, 1976). On the politics of the antinuclear movement in Europe, see Dorothy Nelkin and Michael Pollak,* The Atom Besieged *(Cambridge, Mass., and London: MIT Press, 1981).*

On the Fermi accident of 1966 see John Fuller, We Almost Lost Detroit *(New York: Ballantine, 1975), and the more scholarly E. Pauline Alexanderson, ed.,* Fermi-1, New Age for Nuclear Power *(LaGrange Park, Ill.: American Nuclear Society, 1979). On Three Mile Island, See Philip L. Cantelon and Robert C. Williams,* Crisis Contained: The Department of Energy at Three Mile Island *(Carbondale, Ill.: Southern Illinois University Press, 1982).*

---

68. Atomic Energy Act of 1954 (selections)

"Sec. 103. Commercial Licenses.—

"a. Subsequent to a finding by the Commission as required in section 102, the Commission may issue licenses to transfer or receive in interstate commerce, manufacture, produce, transfer, acquire, possess, import, or export under the terms of an agreement for cooperation arranged pursuant to section 123, such type of utilization or production facility. Such licenses shall be issued in accordance with the provisions of chapter 16 and subject to such conditions as the Commission may by rule or regulation establish to effectuate the purposes and provisions of this Act.

"b. The Commission shall issue such licenses on a nonexclusive basis to persons applying therefor (1) whose proposed activities will serve a useful purpose proportionate to the quantities of special nuclear material or source material to be utilized; (2) who are equipped to observe and who agree to observe such safety standards to protect health and to minimize danger to life or property as the Commission may by rule establish; and (3)

This document appeared in *Laws of 83rd Congress—2nd Session,* August 30, 1954, pp. 1098, 1118–21.

who agree to make available to the Commission such technical information and data concerning activities under such licenses as the Commission may determine necessary to promote the common defense and security and to protect the health and safety of the public. All such information may be used by the Commission only for the purposes of the common defense and security and to protect the health and safety of the public.

"c. Each such license shall be issued for a specified period, as determined by the Commission, depending on the type of activity to be licensed, but not exceeding forty years, and may be renewed upon the expiration of such period.

"d. No license under this section may be given to any person for activities which are not under or within the jurisdiction of the United States, except for the export of production or utilization facilities under terms of an agreement for cooperation arranged pursuant to section 123, or except under the provisions of section 109. No license may be issued to any corporation or other entity if the Commission knows or has reason to believe it is owned, controlled, or dominated by an alien, a foreign corporation, or a foreign government. In any event, no license may be issued to any person within the United States if in the opinion of the Commission, the issuance of a license to such person would be inimical to the common defense and security or to the health and safety of the public.

"Chapter 16. Judicial Review and
Administrative Procedure

"Sec. 181. General.—The provisions of the Administrative Procedure Act (Public Law 404, Seventy-ninth Congress, approved June 11, 1946) shall apply to all agency action taken under this Act, and the terms 'agency' and 'agency action' shall have the meaning specified in the Administrative Procedure Act: Provided, however, that in the case of agency proceedings or actions which involve Restricted Data or defense information, the Commission shall provide by regulation for such parallel procedures as will effectively safeguard and prevent disclosure of Restricted Data or defense information to unauthorized persons with minimum impairment of the procedural rights which would be available if Restricted Data or defense information were not involved.

"Sec. 182. License Applications.—

"a. Each application for a license hereunder shall be in writing and shall specifically state such information as the Commission, by rule or regulation, may determine to be necessary to decide such of the technical and financial qualifications of the applicant, the character of the applicant, the citizenship of the applicant, or any other qualifications of the applicant as the Commission may deem appropriate for the license. In connection with applications for licenses to operate production or utilization facilities, the applicant shall state such technical specifications, including information of the amount, kind, and source of special nuclear material required, the place of the use, the specific characteristics of the facility, and such other information as the Commission may, by rule or regulation, deem necessary in order to enable it to find that the utilization, or production of special nuclear material will be in accord with the common defense and security and will provide adequate protection to the health and safety of the public. Such technical specifications shall be a part of any license issued. The Commission may at any time after the filing of the original application, and before the expiration of the license, require further written statements in order to enable the Commission to determine whether the application should be granted or denied or whether a license should be modified or revoked. All applications and statements shall be signed by the applicant or licensee under oath or affirmation.

"b. The Commission shall not issue any license for a utilization or production facility for the generation of commercial power under section 103, until it has given notice in writing to such regulatory agency as may have jurisdiction over the rates and services of the proposed activity, to municipalities, private utilities, public bodies, and cooperatives within transmission distance authorized to engage in the distribution of electric energy and until it has published notice of such application once each week for four consecutive weeks in the Federal Register, and until four weeks after the last notice.

"c. The Commission, in issuing any license for utilization or production facility for the generation of commercial power under section 103, shall give preferred consideration to applications for such facilities which will be located in high cost power areas in the United States if there are conflicting applications resulting from limited opportunity for such license. Where such conflicting applications resulting from limited opportunity for such license include those submitted by public or cooperative bodies such applications shall be given preferred consideration.

"Sec. 183. Terms of Licenses.—Each license shall be in such form and contain such terms and conditions as the Commission may, by rule or regulation, prescribe to effectuate the provisions of this Act, including the following provisions:

"a. Title to all special nuclear material utilized or produced by facilities pursuant to the license, shall at all times be in the United States.

"b. No right to the special nuclear material shall be conferred by the license except as defined by the license.

"c. Neither the license nor any right under the license shall be assigned or otherwise transferred in violation of the provisions of this Act.

"d. Every license issued under this Act shall be subject to the right of recapture or control reserved by section 108, and to all of the other provisions of this Act, now or hereafter in effect and to all valid rules and regulations of the Commission.

"Sec. 184. Inalienability of Licenses.—No license granted hereunder and no right to utilize or produce special nuclear material granted hereby shall be transferred, assigned or in any manner disposed of, either voluntarily or involuntarily, directly or indirectly, through transfer of control of any license to any person, unless the Commission shall, after securing full information, find that the transfer is in accordance with the provisions of this Act, and shall give its consent in writing. The Commission may give such consent to the creation of a mortgage, pledge, or other lien upon any facility owned or thereafter acquired by a licensee, or upon any leasehold or other interest in such property, and the rights of the creditors so secured may thereafter be enforced by any court subject to rules and regulations established by the Commission to protect public health and safety and promote the common defense and security.

"Sec. 185. Construction Permits.—All applicants for licenses to construct or modify production or utilization facilities shall, if the application is otherwise acceptable to the Commission, be initially granted a construction permit. The construction permit shall state the earliest and latest dates for the completion of the construction or modification. Unless the construction or modification of the facility is completed by the completion date, the construction permit shall expire, and all rights thereunder be forfeited, unless upon good cause shown, the Commission extends the completion date. Upon the completion of the construction or modification of the facility, upon the filing of any additional information needed to bring the original application up to date, and upon finding that the facility authorized has been constructed and will operate in conformity with the application as amended and in conformity with the provisions of this Act and of the rules and regulations of the Commission, and in the absence of any good cause being shown to the Commission why the granting of a license would not be in accordance with the provisions of this Act, the Commission shall thereupon issue a license to the applicant. For all other purposes of this Act, a construction permit is deemed to be a 'license.'

"Sec. 186. Revocation.—

"a. Any license may be revoked for any material false statement in the application or any statement of fact required under section 182, or because of conditions revealed by such application or statement of fact or any report, record, or inspection or other means which would warrant the Commission to refuse to grant a license on an original application, or for failure to construct or operate a facility in accordance with the terms of the construction permit or license or the technical specifications in the application, or for violation of, or failure to observe any of the terms and provisions of this Act or of any regulation of the Commission.

"b. The Commission shall follow the provisions of section 9(b) of the Administrative Procedure Act in revoking any license.

"c. Upon revocation of the license, the Commission may immediately retake possession of all special nuclear material held by the licensee. In cases found by the Commission to be of extreme importance to the national defense and security or to the health and safety of the public, the Commission may recapture any special nuclear material held by the licensee or may enter upon and operate the facility prior to any of the procedures provided under the Administrative Procedure Act. Just compensation shall be paid for the use of the facility.

"Sec. 187. Modification of License.—The terms and conditions of all licenses shall be subject to amendment, revision, or modification, by reason of amendments of this Act or by reason of rules and regulations issued in accordance with the terms of this Act.

"Sec. 188. Continued Operation of Facilities.—Whenever the Commission finds that the public convenience and necessity or the production program of the Commission requires continued operation of a production facility or utilization facility the license for which has been revoked pursuant to section 186, the Commission may, after consultation with the appropriate regulatory agency, State or Federal, having jurisdiction, order that possession be taken of and such facility be operated for such period of time as the public convenience and necessity or the production program of the Commission may, in the judgment of the Commission, require, or until a license for the operation of the facility shall become effective. Just compensation shall be paid for the use of the facility.

"Sec. 189. Hearings and Judicial Review.—

"a. In any proceeding under this Act, for the granting, suspending, revoking, or amending of any license or construction permit, or application to transfer control, and in any proceeding for the issuance or modification of rules and regulations dealing with the activities of licensees, and in any proceeding for the payment of compensation, an award or royalties under sections 153, 157, 186 c., or 188, the Commission shall grant a hearing upon the request of any person whose interest may be affected by the proceeding, and shall admit any such person as a party to such proceeding.

"b. Any final order entered in any proceeding of the kind specified in subsection a. above shall be subject to judicial review in the manner prescribed in the Act of December 29, 1950, as amended (ch. 1189, 64 Stat.

1129), and to the provisions of section 10 of the Administrative Procedure Act as amended."

69. Report of the AEC Advisory Committee on Reactor Safeguards on the Fermi Nuclear Power Plant, 1956

UNITED STATES
ATOMIC ENERGY COMMISSION
WASHINGTON 25, D.C.

June 6, 1956

Mr. K. E. Fields
General Manager
U.S. Atomic Energy Commission
Washington 25, D.C.

SUBJECT: Power Reactor Development Company
(Atomic Power Development Associates)
East Power Reactor

(Reports APDA 108 and 114)*

Dear Mr. Fields:

The present status of the reactor being proposed by the Power Reactor Development Company, associated with Atomic Power Development Associates, was reviewed by the Advisory Committee on Reactor Safeguards at its Eighteenth Meeting on June 3, 1956. This review included the design of the reactor, the state of information on the nuclear properties and the relation of the reactor to its containment and its site. The proposed PRDC reactor represents a greater step beyond the existing state of the art than any other reactor of comparable power level which has been proposed by an industrial group.

From this review the following conclusions were derived:

1. Even though there are no facts or calculations available to the Committee that clearly indicate that the proposed reactor is not safe for this site, the Committee believes there is insufficient information

This document appeared in United States Atomic Energy Commission Records, United States Department of Energy, Germantown, Maryland.

available at this time to give assurance that the PRDC reactor can be operated at this site without public hazard.

2. It appears doubtful that sufficient experimental information will be available in time to give assurance of safe operation of this reactor unless the present fast reactor program of the AEC is amplified and accelerated as detailed below.

3. It is impossible to say whether or not an accelerated program would give sufficient information to permit safe operation of this reactor at the Lagoona Beach site on the time schedule presently proposed.

The following program of investigation is suggested in order to provide information to judge the safety of the proposed operation:

1. The origin of the positive component of the temperature coefficient in EBR-I must be established. A clear demonstration must be given that a coefficient of this magnitude cannot exist in the PRDC design. The experimental program required would involve three phases:

   a. Study of the spontaneous behavior of a new EBR-I core design with rigid fuel elements to insure against the possibility of bowing. Such a study might have to include experiments with both series and parallel flow through core and blanket.

   b. Extensive studies of oscillator experiments on the PRDC design with a simulator, using a wide variety of component temperature coefficients and associated time constants. These studies should be designed to demonstrate that oscillator tests in the startup of the PRDC reactor can produce all the temperature coefficient information required to assure safe transient properties of the reactor, i.e., a negative prompt temperature coefficient of sufficient magnitude to prevent a fuel meltdown.

   c. Further experimental work on CPR-III to show the magnitude and size of the Doppler effect in order to verify the theory.

2. The magnitude, time constant, and sign of the various components of the temperature coefficients in the PRDC design must be evaluated together with a reasonably complete theoretical understanding of their origin in terms of the mechanical design. The program has three aspects:

   a. A demonstration by the simulator studies under 1b.—that the proposed startup program on PRDC can give the information required over a wide range of possible coefficients and time constants.

   b. Conduct of oscillator studies on the EBR-II reactor to show that they are feasible and capable of being interpreted to give the necessary information.

c. Startup program on the PRDC reactor itself to obtain the final information needed before the reactor can be safely operated at full power without meteorological restrictions.

The objective of this program must be to ascertain whether the various negative coefficients are sufficient to prevent meltdown under any conceivable circumstances of control maloperation.

3. The Committee as a whole was not satisfied with the evidence presented that no credible supercriticality accident resulting from meltdown could breach the container. It is felt that a more extensive theoretical and experimental program to examine all the possibilities needs to be established and pursued vigorously. The following are examples: mechanical mock-up studies designed to study distortion of core on sudden melting, criticality studies in ZPR-III design to investigate maximum supercritical arrangements, detailed design studies of the reactor structure, with supporting mock-up experiments, to insure subcritical distribution of melted fuel and to assure that free fall of core parts cannot re-assemble a critical mass suddenly.

4. It is considered critically important that the EBR-II program be pursued more aggressively and coordinated more closely with the PRDC design than is presently the case. The EBR-II program is the only program now constituted which could provide engineering information and operating experience on a high-power-density fast reactor in advance of the scheduled date for operation of the PRDC reactor.

The nature and content of the EBR-II program which the committee considers essential depends on the outcome of investigation 3 above. If it can be shown that a supercritical accident with sufficient energy release to breach the building cannot take place, then the EBR-II program should be aimed at providing general engineering information relevant to the economical design and safe operation of the PRDC reactor.

On the other hand, if it cannot be shown that breaching of the building during a meltdown is impossible, then a much more extensive EBR-II program is required. The test reactor to be operated as EBR-II should then be a genuine prototype of the PRDC reactor. The fuel elements of the test reactor should be identical in all essentials to those proposed for the PRDC reactor, and operated at power densities at least as high as those to be used in the PRDC reactor. The static and dynamic properties of the test reactor should be fully investigated, completely understood theoretically and proved incapable of causing meltdown. These properties should be investigated both for the reactor with its initial charge of U-235 and U-238 and for the reactor with the steady-state concentration of plutonium in the core.

5. The program should not be limited to the above points but should be broadened to whatever extent may be shown necessary by the program itself.

The Committee wishes to note that the experience that now exists on fast power reactors of high power density is not wholly reassuring. While the EBR-I incident is not directly relevant in this connection because the reactor was known to possess an unstable prompt power coefficient under the conditions of the terminal experiment, nevertheless the fact remains that the origin of this unstable coefficient has not been clearly established and therefore its possible occurrence in the PRDC design cannot be definitely excluded on the basis of present experimental information. Opinions differ as to whether its absence can be completely assured in a safe way by the oscillator tests in the pre-startup program proposed for the PRDC reactor *in situ*.

The Committee considers it important that bold steps be taken to advance the development of the fast breeder reactor concept and commends the willingness of the Power Reactor Development Company to risk its capital and prestige in advancing the development of this reactor concept. But the Committee does not feel that the steps to be taken should be so bold as to risk the health and safety of the public. It is important for the AEC to provide sufficient development facilities and experimental information that the safety aspects of the PRDC reactor can be reliably appraised in advance of operation of the reactor itself.

Sincerely yours,
/s /C. Rogers McCullough
C. Rogers McCullough
Chairman
Advisory Committee on
Reactor Safeguards.

*APDA-108   Description of Developmental Fast Neutron Breeder Power Reactor Plant, Sept. 1, 1955.

APDA-114   Location and Environmental Safety of Developmental Fast Breeder Power Reactor Plant.

"Some Supplementary Safeguard Topics Relevant to the Power Reactor Development Company Reactor."

70. AFL-CIO Statement on Fermi, 1956

American Federation of Labor and
Congress of Industrial Organizations
815 Sixteenth Street, N.W.
Washington 6, D.C.

September 12, 1956

Honorable Lewis L. Strauss, Chairman
Atomic Energy Commission
Washington 25, D.C.

Dear Mr. Strauss:

The American Federation of Labor and Congress of Industrial Organizations is extremely disturbed about the recent action of the Atomic Energy Commission in issuing a conditional construction permit for a fast breeder reactor at Lagoona Beach, Monroe County, Michigan. A copy of the statement unanimously adopted by the AFL-CIO Executive Council on August 30 is enclosed.

The Commission has proceeded in this case in a secretive and irregular manner. It has withheld from the public information which properly should have been revealed. It has disregarded and shortcut its own normal procedures for issuing permits and its own safety regulations. The course the Commission has followed in this matter raises serious questions whether it is adequately safeguarding the public interest and the public safety.

Peaceful atomic developments should not proceed under a cover of secrecy. Government censorship in atomic industry should be limited only to that information which must be classified for military security.

Public hearings are required so that all the facts involving the proposed reactor at Lagoona Beach may be opened to public scrutiny before any substantial construction gets underway. Only through such hearings providing interested parties an opportunity to raise the questions which must be answered will gain a full awareness of what is involved in this situation.

Several unions affiliated with the AFL-CIO have filed formal petitions to intervene against the issuance of this construction permit and have requested a full public hearing.

The AFL-CIO strongly urges the Commission that it act favorably on

This document appeared in United States Atomic Energy Commission Records, United States Department of Energy, Germantown, Maryland.

these petitions and that it promptly schedule a public hearing on this most important matter. If such a hearing is ordered, as we trust it will be, we respectfully request an opportunity to participate therein on behalf of our 15,000,000 members and their families.

Sincerely yours,
/s /George Meany
President

Attachment

71. Preliminary Report on Fuel Damage in Fermi Reactor, 1966

It is believed that significant damage has occurred to some of the fuel in the Enrico Fermi reactor. This report summarizes the operations during which this damage appears to have occurred and presents the status of knowledge as of 8 A.M., Monday, October 10, 1966. A more detailed report is being prepared.

On Tuesday, October 4, 1966, reactor operations at the Fermi plant were resumed after a lengthy outage due to the necessity for repairs to the No. 1 and No. 3 steam generators. The reactor was kept at 1 Mw during the night of October 4–5 during which time reactor heat was used to raise the reactor temperature to the normal level for power start-up.

At 1:45 P.M. on Wednesday, October 5, following a number of minor delays due to adjustments in the feedwater control systems, an increase of the reactor power level was initiated at the normal rate. Two primary and secondary coolant loops were in operation. By 3 P.M., the power level had been raised to 34 Mw(th), a level which had been surpassed many times in the past. At about this time the staff member present observed that the control rods seemed to be somewhat further out of the core than would be normal under the circumstances. He went to the back of the control panel to observe the outlet temperatures from those core subassemblies which have thermocouples at their outlets. He saw that two core subassembly outlet temperatures were higher than normal. At about the same time the

This document appeared in United States Atomic Energy Commission Records, United States Department of Energy, Germantown, Maryland.

fission product detector indicated a substantial increase in the primary cover gas activity, and the reactor building radiation monitors indicated a rise in the radiation level in the building large enough so that the reactor building was automatically isolated. The radiation monitor charts on the waste gas stack show that essentially no radioactivity was released to the atmosphere.

The reactor power was reduced manually from 34 to 29 Mw upon the rise in the cover gas activity, and the reactor was scrammed shortly after the building isolation occurred. The scram rods operated normally in shutting the reactor down. The radiation levels in the reactor building were not in themselves dangerously high and no personnel over-exposure occurred. All health physics procedures were carried out according to the prepared written instructions.

A meeting was held at the plant shortly after the occurrence to determine whether any measures needed to be taken either for public or personnel safety or for protection of plant equipment. No such measures were required but operations were initiated to obtain samples of the primary sodium and the primary cover gas. The AEC, Division of Compliance, was notified of the occurrence by telephone immediately following the meeting.

A number of facts have been developed since the reactor was shut down on October 5. From the standpoint of significance in indicating the nature and extent of fuel damage the following are the most important:

1. Analysis of the primary cover gas indicates concentrations of fission gases, notably Xe-133 and Xe-135, which would be characteristic of concentrations resulting from the release of fission products from an amount of fuel about equal to that in a core subassembly.
2. Analysis of the reactor building atmosphere indicates the presence of the same fission product gases in about the concentrations which would be expected due to the known leak rate from the primary cover gas system to the reactor building. No isotopes of iodine were determinable above background in the gamma-ray spectrometer used for the analysis.
3. A subcritical reactivity check was made on October 7 which indicated that the core reactivity was about 22 cents lower than when the reactor was brought critical on October 4.
4. Correlation of reactor control rod position, power level, and time, indicates that reactivity began to be lost on October 5 when the power level reached about 15 Mw(th). Loss of reactivity was roughly linear with power thereafter.

5. Allowing for the 12 minute over-all time constant of the fission product detector, the first indication of fission product release agrees closely with the first loss of reactivity.

6. The readings on the two core subassembly outlet thermocouples indicated that the temperature difference across these subassemblies was considerably greater than normal.

Taken together these facts indicate that fuel damage has taken place, and that some fuel redistribution has taken place. There is a possibility that the damage is limited to the two core subassemblies which indicated high outlet temperatures but, since only one fourth of the subassemblies have thermocouples, this would be fortuitous.

The reactor is presently shut down and the activity in the reactor building atmosphere is low enough to allow accessibility. No surface activity has been found anywhere in the reactor building. All fission products except some of the noble gases and their daughters appear to have been retained in the primary system.

---

72. Calvert Cliffs Coordinating Committee versus U.S. Atomic Energy Committee, 1971

CALVERT CLIFFS COORDINATING
COMMITTEE, INC., et al,
Petitioners,

v.

UNITED STATES ATOMIC ENERGY
COMMISSION and United States
of America, Respondents.
Nos. 24839, 24871.
United States Court of Appeals,
District of Columbia Circuit.
Argued April 16, 1971.
Decided July 23, 1971.

Proceeding to review order of Atomic Energy Commission. The Court of Appeals, J. Skelly Wright, Circuit Judge, held that the courts have

This document appeared in Calvert Cliffs Coordinating Committee, Inc., et al., Petitioners, v. USAEC and USA, Respondents, July 23, 1971; *449F. 2d 1109 (1971)*. Federal Reporter Series, pp. 1111–31 passim.

power to require agencies to comply with procedural directions of National
Environmental Policy Act of 1969 and that the Commission's rules precluding review consideration of nonradiological evironmental issues unless specifically raised, prohibiting raising such issues in certain pending proceedings or when issues have been passed on by other agencies, and precluding
consideration between grant of construction permit and consideration of
grant of operating license, did not comply with Act.

Remanded . . .

J. Skelly Wright, Circuit Judge:

These cases are only the beginning of what promises to become a flood
of new litigation—litigation seeking judicial assistance in protecting our
natural environment. Several recently enacted statutes attest to the commitment of the Government to control, at long last, the destructive engine
of material "progress." But it remains to be seen whether the promise of
this legislation will become a reality. Therein lies the judicial role. In these
cases, we must for the first time interpret the broadest and perhaps most
important of the recent statutes: the National Environmental Policy Act of
1969 (NEPA). We must assess claims that one of the agencies charged with
its administration has failed to live up to the congressional mandate. Our
duty, in short, is to see that important legislative purposes, heralded in the
hall of Congress, are not lost or misdirected in the vast hallways of the
federal bureaucracy.

NEPA, like so much other reform legislation of the last 40 years, is cast
in terms of a general mandate and broad delegation of authority to new and
old administrative agencies. It takes the major step of requiring all federal
agencies to consider values of environmental preservation in their spheres
of activity, and it prescribes certain procedural measures to ensure that
those values are in fact fully respected. Petitioners argue that rules recently
adopted by the Atomic Energy Commission to govern consideration of environmental matters fail to satisfy the rigor demanded by NEPA. The Commission, on the other hand, contends that the vagueness of the NEPA
mandate and delegation leaves much room for discretion and that the rule
challenged by petitioners falls well within the broad scope of the Act. We
find the policies embodied in NEPA to be a good deal clearer and more
demanding than does the Commission. We conclude that the Commission's procedural rules do not comply with the congressional policy. Hence
we remand these cases for further rule making.

The special importance of the pre-operating license stage is not difficult
to fathom. In cases where environmental costs were not considered in
granting a construction permit, it is very likely that the planned facility will
include some features which do significant damage to the environment and

which could not have survived a rigorous balancing of costs and benefits. At the later operating license proceedings, this environmental damage will have to be fully considered. But by that time the situation will have changed radically. Once a facility has been completely constructed, the economic cost of any alteration may be very great. In the language of NEPA, there is likely to be an "irreversible and irretrievable commitment of resources," which will inevitably restrict the Commission's options. Either the licensee will have to undergo a major expense in making alterations in a completed facility or the environmental harm will have to be tolerated. It is all too probable that the latter result would come to pass.

By refusing to consider requirement of alterations until construction is completed, the Commission may effectively foreclose the environmental protection desired by Congress. It may also foreclose rigorous consideration of environmental factors at the eventual operating license proceedings. If "irreversible and irretrievable commitment(s) of resources" have already been made, the license hearing (and any public intervention therein) may become a hollow exercise. This hardly amounts to consideration of environmental values "to the fullest extent possible."

A full NEPA consideration of alterations in the original plans of a facility, then, is both important and appropriate well before the operating license proceedings. It is not duplicative if environmental issues were not considered in granting the construction permit. And it need not be duplicated, absent new information or new developments, at the operating license stage. In order that the pre-operating license review be as effective as possible, the Commission should consider very seriously the requirement of a temporary halt in construction pending its review and the "backfitting" of technological innovation. For no action which might minimize environmental damage may be dismissed out of hand. Of course, final operation of the facility may be delayed thereby. But some delay is inherent whenever the NEPA consideration is conducted—whether before or at the license proceedings. It is far more consistent with the purposes of the Act to delay operating at a stage where real environmental protection may come about than at a stage where corrective action may be so costly as to be impossible.

(12) Thus we conclude that the Commission must go farther than it has in its present rules. It must consider action, as well as file reports and papers, at the pre-operating license stage. As the Commission candidly admits, such consideration does not amount to a retroactive application of NEPA. Although the projects in question may have been commenced and initially approved before January 1, 1970, the Act clearly applies to them since they must still pass muster before going into full operation. All we demand is that the environmental review be as full and fruitful as possible.

73. Three Mile Island; *Nuclear News* Special Report, April 6, 1979

A combination of design deficiency, mechanical failure, and human error contributed to the ill-controlled accident that was touched off at about 4 A.M. on Wednesday, March 28, at Unit 2 of the Three Mile Island nuclear power station of Metropolitan Edison Company, a member company of General Public Utilities (GPU). The initial event at the unit, located near Harrisburg, Pa., has been characterized as a loss-of-normal-feedwater turbine trip—with complications. As Norman Rasmussen of M.I.T. explained the following Sunday on ABC's television program "Issues and Answers," the event could not be considered, in the parlance of reactor safety studies, a "major event" ("an event of major consequences" to the public health and safety, he quickly went on to clarify). The TMI-2 "event," however, certainly promises to have major consequences for the utilization of nuclear power in America and elsewhere—this in spite of the fact that the accident was contained and the amounts and forms of radioactivity that did escape from the plant, appeared, by most accounts, to have been of no major consequence.

If not calamitous, however, the accident was indeed most serious. Considerable fuel failure occurred, fission products were released into and out of the containment and into the atmosphere, and a tremendous loss, in terms of failed and suspect fuel, the staggering costs of eventual cleanup, and lost confidence, has been visited upon Metropolitan Edison—not to speak of the entire nuclear effort.

Buildup to a Crisis

The TMI-2 incident occurred one year to the day after it had first achieved criticality. It was operating at 98 percent of full power. In a matter of minutes during the early morning hours of March 28, it was off line, the secondary system was out of service, the pressurizer relief valve was stuck open, and primary system water was spilling into the containment structure, subsequently to be pumped out to the reactor auxiliary building, from which radioactive gases would escape to the atmosphere (see "How It Happened").

In spite of the severity of the situation, it was not until almost 7 A.M. that a site emergency was declared. High activity readings in the reactor building were then noted, prompting the call for a general emergency at about 7:30 A.M. The utility also notified the Nuclear Regulatory Commis-

This essay appeared in *Nuclear News*, Special Report, April 6, 1979.

sion's Region I office outside Philadelphia, and Civil Defense in nearby Harrisburg. State and federal officials alike have taken Metro Ed to task for not issuing notification until nearly four hours after the initiation of the accident.

The NRC dispatched a team to the site and began taking radiation readings—as did the Pennsylvania Department of Environmental Resources. Ventilation in the auxiliary building continued, radioactivity in the containment remained high, and the "bubble" in the reactor presented a major obstacle to achievement of cold shutdown. By the end of Wednesday, it was generally agreed that this was the most severe commercial nuclear accident to date, and that the worst may not have been over.

As the gravity of the situation became clear, the NRC beefed up its presence at the site. Harold Denton, director of the Office of Nuclear Reactor Regulation, led a contingent from NRC headquarters and became the agency's spokesman at the frequent news conferences held during the next few days. Help was also brought in from other NRC regional offices, and from virtually every national laboratory.

Also sent in were utility and industry teams (including, of course, the B&W personnel), teams from state and federal agencies, and scientists and engineers from outside the United States. Other teams in Washington, D.C., and elsewhere in the United States worked full-time on the problem.

A huge press contingent also gathered in the towns of Middletown and Goldsboro very close to the site, and in Harrisburg. The press generally is credited with doing a good job of reporting on the incident in spite of some confusion during the initial states.

There was often disagreement in the reports from the Metropolitan Edison Company and those of the NRC concerning the status of the plant and the progress in the efforts to cool down the reactor. Some of this was understandable as it was apparently not known at first how serious the problem was and, in some instances, because one source had more recent information than the other. But even when working from similar information, the reports from the company and NRC differed, with the reports from the utility almost always carrying a more optimistic tone than those from the Commission. As the press increasingly noted these differences, and as the situation was becoming acrimonious between the NRC and the utility, the White House ordered on April 1 that all further technical reports on the status of the plant were to come from the Commission.

Radioactivity

By the morning of April 3, the NRC had announced the following cumulative data from environmental monitoring:

• No radioiodine was detected in any of the 130 water samples taken by the NRC, the Department of Energy, and the Commonwealth of Pennsylvania since the accident. From what was calculated as the amount of radioiodine released from the plant by water, the thyroid dose to anyone drinking the water was estimated to be less than 0.2 millirem.

• Eight of the 152 offsite air samples showed radioiodine present, the most activity measured being 2.4×10 microcuries per millilitre. Thyroid dose to anyone at the site boundary was estimated to be less than 50 mrem over a five-day period. Radioactive inert gases were detected in the sample, taken within a 65-km radius of the plant, but the maximum activity detected was about one-fourth of the permissible concentration established in 10CFR20.

• Milk samples tested by the state did not yield radioiodine, but the federal Food and Drug Administration found results ranging from 14 to 40 picocuries of I-131 per litre in the nine samples it took. Herds are to be placed on stored feed, according to the Department of Health, Education and Welfare, when the I-131 count reaches 12,000 pCi/l. The thyroid dose for anyone drinking milk is estimated to be less than 0.5 mrem per day.

• No radioiodine was detected in any of 147 soil samples taken by the NRC and the DOE, nor in 171 vegetation samples taken within 3 km of the site by the NRC, the DOE, and the state.

Radioiodine in the plant was determined (activity adjusted to decay as of midnight, April 2–3) to be 5.1×10^6 Ci in the core, 3.32×10^6 Ci in the coolant, and 0.235 Ci in the industrial waste treatment system; in all cases, I-131 accounts for at least 96 percent of the activity, and I-133 for the remainder. The industrial waste treatment system was continuing to be filled through the turbine building sump, and capacity was limited. Releases from this system to the Susquehanna River since the accident have stirred much public resentment.

Three plant employees were exposed to radiation beyond the 3 rem per quarter level by the end of the quarterly monitoring period on March 31. Dosages of 3.1, 3.4, and 3.8 rem were received when the workers tried to drain and secure the auxiliary building. At least 12 other workers received considerable doses below 3 rem.

Hitting Home

Technical aspects aside, what established this as the most serious nuclear accident was its effect upon the daily lives of the public. For the first time, a problem in a nuclear power plant disrupted the personal and business activities of many thousands of people.

Initially, most of the residents in the area went about their daily routines, but this quickly changed. While some residents did continue to state they felt they were in no great danger, many others expressed feelings ranging from anger at the inconvenience caused by the situation to fear for their health and safety. Widespread distrust of the statements by utility spokesmen was reported. "I just don't believe them" was a typical response. Others expressed concerns for their property values and their businesses.

On Friday, March 30, Gov. Richard Thornburgh ordered several schools closed in the area of the plant and recommended the evacuation of preschool children and pregnant women within five miles of the plant. Also, he urged people living within 10 miles to remain indoors. These actions were taken on the recommendation of NRC Chairman Joseph Hendrie and following an "unplanned and unexpected release" of radioactive gas. The following day, Thornburgh said it was no longer seen as necessary that people living within 10 miles of the plant remain indoors, but he still advised that pregnant women and preschool children move from the area five miles around the plant.

Also, plans were prepared for a general evacuation for everyone out to a distance of 20 miles from the plant. Thornburgh announced that a precautionary evacuation might be ordered should the efforts in bringing the reactor to a cold shutdown state be considered sufficiently risky to require evacuation. Such an evacuation was most discussed on March 30–31 and April 1, when the hydrogen "bubble" in the reactor was complicating efforts to cool down the core.

On Sunday, April 1, President Carter visited the site, with the reason for his visit interpreted widely as that of reassuring residents while preparing them for a general evacuation, if that became necessary. If it did, he said, the evacuation instructions should be followed "calmly and exactly." By April 2, definite progress was noted in removing the hydrogen bubble, and Thornburgh announced that a general evacuation would probably not be needed.

Estimates varied greatly about the number of people who moved out of the area, but civil defense officials said that of the 200,000 people living within 20 miles of the plant, an estimated 80,000 left over the weekend of March 31–April 1; 40,000 were still away on Tuesday, April 3; and 30,000 to 35,000 had not yet returned on April 5. In addition, an undetermined number of businesses were closed.

Helping somewhat to alleviate the disgruntlement of the area people was the presence of American Nuclear Insurers, which set up an emergency claims office in Harrisburg. ANI staff members distributed checks to evacuees to cover their hotel and meal expenses, and by Tuesday, April 3, about $200,000 had been paid out. A spokesman from ANI said that the pool of nuclear insurance firms had "no idea" how much would be paid out in such claims, and that ANI had "no estimates on either side" on property and liability claims. Prior to the TMI-2 situation, ANI had paid a total of only $620,000 in claims since the insurance pools were formed in 1957.

On April 3 and 4, with the situation improving at the plant, and with people returning and some schools and businesses reopening, the area was starting to return to some semblance of normalcy. Still to be determined were just how permanent would be the loss of faith by the area residents in the plant's safety, and any adverse impact on the local economy.

The accident sharply changed the plans of the whole General Public Utilities Corporation group (which also includes Jersey Central Power and Light Company and Pennsylvania Electric Company). Even if the outcome is the most favorable for the utility—i.e., if TMI-2 can be repaired and returned to service at a tolerable cost—the expenditure in money and personnel will be sizable. The situation will not be helped if TMI-1, which was off line for routine maintenance at the time of the TMI-2 accident, is forced to remain out of service very long; TMI-1 has been a dependable, highly productive unit during its four and a half years of operation, and a prolonged outage will add replacement power costs to GPU's other worries.

Aware of its predicament, GPU moved on April 3 to suspend all construction work on new power plants (including the Forked River nuclear unit) and all but the most essential construction on existing facilities. Non-essential maintenance will be reduced or postponed wherever possible, and scheduled outages may be postponed.

Political Reactions

At the political level, reactions came quickly, and, not unexpectedly, there were few to take up any defense of nuclear power. On Thursday, March 29, the morning after the accident, all five NRC commissioners, along with top NRC aides, were called before the House Interior Committee's Subcommittee on Energy and the Environment (Rep. Morris K. Udall [D., Ariz.], chairman) to brief the committee on the accident. Reflecting the obvious concern of their constituents, the Congressmen were at times sharp in their questioning of NRC Chairman Hendrie and his colleagues. Hendrie was asked by Rep. James Weaver (D., Ore.), "How close did we come to the China Syndrome?" Hendrie replied, "We were nowhere near it, nowhere near it." The Congressmen persisted, "*Could* it happen?" Hendrie answered, "In principle, yes," but he insisted that the chances that it would happen were very small. Weaver indicated he was not assured by the NRC Chairman's responses.

Representative Udall did not hide his skepticism concerning the safety of nuclear power either at the hearing or on the same "Issues and Answers" program alluded to earlier, where he urged that America's reliance on nuclear power for the future be seriously questioned. He said that since the country is now dependent on this source for no more than 10–12 percent of its electricity, it would not be too difficult to curtail future efforts in this direction and to shift to other alternatives. He said the people had been told that this type of accident would not happen, and now it has happened. Rasmussen later corrected this by saying that this type of event was among those covered in the Reactor Safety Study (WASH-1400). Rasmussen agreed with Udall that a thorough review of similar reactors would be necessary to prevent similar occurrences.

The NRC, in the meantime, has directed the operators of all nine Babcock & Wilcox reactors having operating licenses to review their facilities and procedures in light of the TMI-2 incident (see "Other B&W Units Cleared to Operate"). The Commission also, as part of its investigation of the incident, is setting up a special review group similar to the one established in the case of the Brown's Ferry fire.

Over and above these efforts by the NRC, Pennsylvania's Sen. Richard Schweiker (R.) has called for the formation of a Presidential commission to analyze "the full implications of the Three Mile Island accident" and has asked that this analysis include what role nuclear power should play in the future of U.S. electricity generation.

Other legislation doubtless will be generated from the hearings being conducted by various committees on both sides of Congress. Sen. Gary Hart (D., Colo.), chairman of the Subcommittee on Nuclear Regulation, said he plans to propose legislation that would strengthen federal oversight of all nuclear power plants. His legislation would put the government, rather than the utility, clearly in charge if and when a serious accident should occur. Hart suggested that all plants be monitored remotely by the NRC, possibly by satellite communication.

Somewhat along those same lines, President Carter signed almost immediately after the accident an executive order centralizing much of the federal emergency planning in a single agency, the Federal Emergency Management Agency. This agency combines the Defense Civil Preparedness Agency, the Federal Disaster Assistance Administration, the Federal Preparedness Agency, the Federal Insurance Administration, and the National Fire and Control Administration.

Actions were taken, too, at the state level. California Gov. Jerry Brown sent a telegram to the NRC after the accident urging that the Rancho Seco nuclear plant near Sacramento, a B&W unit, be ordered shut down until the Three Mile Island incident was fully studied and corrective measures could be taken.

A few brave souls in the public eye did make statements in defense of nuclear power—notably, Department of Energy Secretary James Schlesinger, who two days after the accident cited the "historic record" of nuclear power as "excellent" and urged that the accident be viewed in proper perspective. He noted that the radiation exposures experienced were "very limited" and contended that nuclear power must continue to be "an essential element" in the nation's drive for energy independence. Sen. J. Bennett Johnston (D., La.) was quoted soon after the accident as saying that it "could lead to an agonizing reappraisal of our use of nuclear power," but adding that "if you're going to have a viable economy, you've simply got no choice but to have nuclear power." The Senator's ambivalence clearly describes the quandary in which many public officials—and the American people—now find themselves with respect to nuclear power.

A President Treads Lightly

In his energy speech on April 5—one week after the Three Mile Island accident—President Carter had little to say about the future role of nuclear power in achieving the stated goal of energy sufficiency. Just as he wore protective plastic shoe covers in his postaccident visit to TMI-2, the President trod lightly over the subject, saying at the outset of his speech that the accident "demonstrated dramatically that we have *other* energy problems" and adding later that he had established an independent Presidential commission of experts "to investigate the causes of the accident and to make recommendations on how we can improve the safety of nuclear power plants. You deserve a full accounting and you will get it."

Among the avenues toward energy independence, he mentioned solar energy, shale oil, advanced coal technology, improved automobile efficiency, improved mass transit systems, and even wood-burning stoves. He made no mention, however, of nuclear power, whether fission or fusion.

Other B&W Units Cleared to Operate

Directed by the Nuclear Regulatory Commission to review their operations in light of the TMI-2 incident and to report within 10 days on the status of their plants and procedures, the operators of eight other Babcock

& Wilcox reactors will be allowed to continue operating, this being judged "without undue risk to the public," as stated by Edson Case, NRC deputy director of nuclear reactor regulation at a public briefing on April 4. Case noted that the NRC would take whatever action might be necessary after the reports are received.

Of the eight plants, roughly the same but differing slightly (some have supplemental safety systems not found at TMI-2), only five are operating currently. These are: Rancho Seco, operated by Sacramento Municipal Utility District; the three Oconee units of Duke Power Company; and Crystal River-3 (Florida Power Corporation), scheduled for an outage later this month. The other three units are presently down for scheduled outages. These are: TMI-1; Unit 1 of Arkansas Nuclear One (Arkansas P&L); and Davis-Besse-1 (Central Area Power Coordination Group).

Addendum . . . How It Happened

The actual sequence of events beginning early in the morning of March 28 is believed to have gone as follows:

• At about 4 A.M., the main feedwater system of Three Mile Island-2 malfunctioned, apparently as a result of failure either of the demineralizer or of the air supply to an air-operated valve.
• With the main feedwater supply system out of operation the auxiliary feedwater system was to have started automatically. It did not, however, because a number of manual valves in the auxiliary system had inadvertently been left closed after a test of the system in the days prior to the accident.
• Without feedwater supply, the steam generators dried out, resulting in a rise in the primary coolant temperature and pressure. The turbine generator tripped. The overpressurization generated a signal that scrammed the reactor.
• Within seconds, primary loop pressure rose to 1.625X10 Pascal (2355 psi), triggering a relief valve at the top of the pressurizer. The primary coolant, thus relieved, was piped to a quench tank, located within the containment. The relief valve failed in the open position, providing a continuous release of coolant from the system. This failure and the failure in the main feedwater system (both of a mechanical nature), and the malfunction of the auxiliary feedwater system were the main initiators of the incident.
• The emergency core-cooling system started automatically at two minutes into the accident sequence, and began to raise the coolant level. Within a

few minutes, the level indicator for the pressurizer is reported to have started giving erroneously high readings and in fact went off scale on the high side—while actually the opposite condition should have been indicated. An operator manually overrode first one high-pressure injection pump (ECCS) and, a few minutes later, the other. It is believed that this was done to prevent the pressurizer from becoming "solid" (filled entirely with water); if the pressurizer were filled, pressure control would be lost.

• Within two minutes of the first ECCS shutoff, it has been reported, primary coolant was flashing to steam in the reactor.

• Also after the first ECCS shutoff, containment basement sump pumps began to operate automatically (containment isolation at TMI-2 is not automatic with ECCS initiation, but only after pressure within the containment reaches four psig). Several minutes later the feedwater flow was restarted, increasing pressure in the steam generators.

• With flow through the pressurizer relief valve unchecked, the quench tank continued to fill, and some fifteen minutes into the accident sequence its rupture disc blew open, releasing ultimately upwards of 40,000 litres of primary coolant onto the containment floor. From here the water was pumped into a storage tank in the auxiliary building, where water spilled again after the tank, already partly filled, exceeded capacity. Radioactive isotopes of krypton, iodine, and xenon began to escape to the environment through the building's ventilation system; vent filters trapped much of the iodine but did not stop the noble gases.

• The primary coolant continued to be released from the primary system through the open pressurizer relief valve, resulting in a decrease in primary pressure. After one hour and fifteen minutes after the initial transient, and at about 1300 psi, two primary coolant pumps on one steam generator loop were manually tripped, apparently to prevent cavitation and resultant damage to the pumps. The other two pumps were tripped, again manually, 25 minutes later. At this point the reactor system was without forced convection and without a heat sink.

• Control of the pressurizer relief system was regained, and the ECCS turned on again. It is believed, however, that coolant level had dropped so low that part of the core had been uncovered resulting in substantial fuel damage. Although temperatures in the core registered quite high, no actual melting is believed to have occurred.

• At the three-hour point it was noted that primary-to-secondary leaks in the "B" steam generator had occurred, providing a pathway for more radioactive material to the environment from the secondary system. The "B" steam generator was subsequently isolated; the "A" steam generator remained functional.

• It is postulated that the substantial core damage, combined with high fuel temperatures, resulted in the formation within the reactor vessel of a gas bubble, consisting of hydrogen (from Zircaloy-water reaction) and gaseous fission products. The bubble presented a problem in cooldown: If primary system pressure were reduced, the bubble would expand and possibly impair the cooling mode (e.g., cause pump cavitation).

• Hydrogen also became a problem in the containment. In the beginning phases of the accident the containment pressure had risen above atmospheric. Later on, hydrogen was detected, which, if allowed to accumulate, had the potential for leading to an explosion.

• By Wednesday evening, cooling of the damaged core was restored by activation of one of the primary coolant pumps on the "A" loop, using the "A" steam generator as a heat sink. The degasifying process involved drawing dissolved gas out of the coolant from the cold leg of the primary system and spraying it through the pressurizer and out into the containment. This eventually reduced the size of the bubble but added to the hydrogen in the containment. By April 3, a hydrogen recombiner was in place outside the containment, and hydrogen concentration dropped gradually (2.3 percent, late P.M., April 1; 2.1 percent, early A.M., April 4). It was expected that the concentration would be brought down to 1 percent by about April 15.

• At press time, some coolant temperatures above the core were still in excess of 200 degrees C, requiring primary pressure to be maintained at about 6.9×10 Pa (1000 psi). The approach to cold shutdown that was preferred as of April 6 was to continue cooling with pumped circulation until about April 11, and then convert to "natural circulation" in the primary system and solid water in the secondary. Use of the residual heat removal system (RHRS) was not actively being considered.

X

Afterword

We live in a world where nuclear weapons can literally mean the end of history. History is not so much what has happened as what is remembered, and the destruction of our species would thus eliminate the past in entirety. In these pages we have considered the history of the American atom and thereby the process whereby we have achieved the terrible power that was only imagined in the apocalyptic premonitions of our ancestors: the power of self-destruction and the ability to end history.

In the meantime we muddle through. Living with nuclear energy involves both technology and human choice, necessity and freedom. The necessity involves three fairly intractable realities with which we must live: first, that nuclear weapons exist in the tens of thousands and nuclear reactors in the hundreds worldwide; second, that the human community continues to be organized in units of nation-states according to the principle of national sovereignty; third, that Soviet-American relations are founded on a doctrine of mutual deterrence aimed at avoiding nuclear war.

From this perspective, we are unlikely to eliminate either the nuclear technology we have inherited or the human political framework and super-power relationship that currently exist. We are likely to have to live with and understand these realities to control them. Some choices have been made for us; others still remain.

The American atom is history, an inherited dilemma—how to reconcile national sovereignty and nuclear technology in a manner that allows us to control the dangerous machines that we have historically considered to be essential to our national security and productive well being. Nuclear energy is not the product of a conspiracy, despite its legacy of secrecy. In 1945 an overwhelming majority of Americans supported dropping the first nuclear weapons on Japan; in the 1950s and 1960s a majority supported the development of nuclear power to satisfy the national appetite for electrical energy and its attendant home conveniences.

Thus we may object to the American atom, but we cannot deny our historical choice to benefit from its power in time of both war and peace. The purpose of this volume has been to delineate how that choice was made and what its consequences have been, so that we may choose wisely in a technological future of both risks and benefits. The American atom is history. It is also the greatest challenge to our future survival as a species on this planet. We must master its past if we are to control its future.

Index

accidents. *See* nuclear weapons

Acheson, Dean, 72, 73

Acheson-Lilienthal Report, 72, 150

Administrative Procedure Act, 300, 301; and United States Atomic Energy Commission, 297

air-to-surface ballistic missiles (ASBMs), 275, 276, 277, 280

Alamogordo (N.M.), 50, 54, 72, 179; and Trinity test, 25, 47, 48, 178

Albuquerque, and Trinity test, 48, 50

Allison, Samuel, 30

American Federation of Labor and Congress of Industrial Organizations: Executive Council, 305; and Fermi reactor accident, 305–6

American Nuclear Insurers, and Three Mile Island accident, 314–15

Anderson, Oscar E., Jr., 26, 74

Anglo-American Declaration of Trust, 43

Antarctica, as non-nuclear zone, 238

anti-ballistic missiles (ABMs), 212, 239, 261, 272; treaty limiting, 260–65, 272, 274

Argonne National Laboratory (Chicago), 99, 122

Arkansas Nuclear One Reactor, and Babcock and Wilcox, 318

Arkansas Power and Light Company, 318

arms control, 234, 238–39, 289

arms race, 76, 205, 227, 248

Arnold, Gen. Henry H., 47

Articles of Agreement Governing Collaboration Between The Authorities of the USA and the UK in the Matter of Tube Alloys. *See* Quebec Agreement

atomic bomb, 2, 10, 13, 16, 18, 20, 20–21, 22, 24, 25, 26, 40, 45, 48, 67, 68, 75–76, 77, 84, 85, 93, 95, 104, 105–6, 107, 123, 124, 129, 130, 131, 132, 139, 141, 160, 167, 172, 174, 213, 214, 215, 216, 217, 218; and Russians, 76; test of Joe I, 114; first Soviet, 115–16, 116–17, 128, 151; and J. Robert Oppenheimer, 163

Atomic Development Authority, and Baruch Plan, 94, 95, 96

atomic energy, 8, 19, 110, 212; postwar control of, 71, 72; federally conducted research and development, 79; private research and development, 79; peaceful uses of, 80

Atomic Energy Act of 1946 (McMahon Act), 7, 79–92, 114, 154, 163, 293, 294; foundation of U.S. Atomic Energy Commission, 80; to assist research, 80–81; and federal atomic research, 81; and production of fissionable materials, 81–82; control of fissionable materials, 82–85; and military applications of atomic power, 84–85; and atomic energy devices, 85–86; and property of Commission, 86–87; dissemination of information, 87–88; and patents, 88–89; organizational and general authority of,